The Mason-Bees

J. Henri Fabre

Translated by Alexander Teixeira de Mattos

TRANSLATOR'S NOTE.

This volume contains all the essays on the Chalicodomae, or Mason-bees proper, which so greatly enhance the interest of the early volumes of the "Souvenirs entomologiques. " I have also included an essay on the author's Cats and one on Red Ants—the only study of Ants comprised in the "Souvenirs"—both of which bear upon the sense of direction possessed by the Bees. Those treating of the Osmiae, who are also Mason-Bees, although not usually known by that name, will be found in a separate volume, which I have called "Bramble-bees and Others" and in which I have collected all that Fabre has written on such other Wild Bees as the Megachiles, or Leaf-cutters, the Cotton-bees, the Resin-bees and the Halicti.

The essays entitled "The Mason-bees, Experiments" and "Exchanging the Nests" form the last three chapters of "Insect Life", translated by the author of "Mademoiselle Mori" and published by Messrs. Macmillan, who, with the greatest courtesy and kindness have given me their permission to include a new translation of these chapters in the present volume. They did so without fee or consideration of any kind, merely on my representation that it would be a great pity if this uniform edition of Fabre's Works should be rendered incomplete because certain essays formed part of volumes of extracts previously published in this country. Their generosity is almost unparalleled in my experience; and I wish to thank them publicly for it in the name of the author, of the French publishers and of the English and American publishers, as well as in my own.

Some of the chapters have appeared in England in the "Daily Mail", the "Fortnightly Review" and the "English Review"; some in America in "Good Housekeeping" and the "Youth's Companion"; others now see the light in English for the first time.

I have again to thank Miss Frances Rodwell for the invaluable assistance which she has given me in the work of translation and in the less interesting and more tedious department of research.

ALEXANDER TEIXEIRA DE MATTOS.

Chelsea, 1914.

CONTENTS.

CHAPTER 1
THE MASON-BEES.

Reaumur (Rene Antoine Ferchault de Reaumur (1683-1757), inventor
of the Reaumur thermometer and author of "Memoires pour servir a
l'histoire naturelle des insectes. " —Translator's Note.) devoted one
of his papers to the story of the Chalicodoma of the Walls, whom he
calls the Mason-bee. I propose to go on with the story, to complete it
and especially to consider it from a point of view wholly neglected
by that eminent observer. And, first of all, I am tempted to tell how I
made this Bee's acquaintance.

It was when I first began to teach, about 1843. I had left the normal
school at Vaucluse some months before, with my diploma and all the
simple enthusiasm of my eighteen years, and had been sent to
Carpentras, there to manage the primary school attached to the
college. It was a strange school, upon my word, notwithstanding its
pompous title of 'upper'; a sort of huge cellar oozing with the
perpetual damp engendered by a well backing on it in the street
outside. For light there was the open door, when the weather
permitted, and a narrow prison-window, with iron bars and lozenge
panes set in lead. By way of benches there was a plank fastened to
the wall all round the room, while in the middle was a chair bereft of
its straw, a black-board and a stick of chalk.

Morning and evening, at the sound of the bell, there came rushing in
some fifty young imps who, having shown themselves hopeless
dunces with their Cornelius Nepos, had been relegated, in the phrase
of the day, to 'a few good years of French. ' Those who had found
mensa too much for them came to me to get a smattering of
grammar. Children and strapping lads were there, mixed up
together, at very different educational stages, but all incorrigibly
agreed to play tricks upon the master, the boy master who was no
older than some of them, or even younger.

To the little ones I gave their first lessons in reading; the intermediate
ones I showed how they should hold their pen to write a few lines of
dictation on their knees; to the big ones I revealed the secrets of
fractions and even the mysteries of Euclid. And to keep this restless
crowd in order, to give each mind work in accordance with its
strength, to keep attention aroused and lastly to expel dullness from
the gloomy room, whose walls dripped melancholy even more than

dampness, my one resource was my tongue, my one weapon my stick of chalk.

For that matter, there was the same contempt in the other classes for all that was not Latin or Greek. One instance will be enough to show how things then stood with the teaching of physics, the science which occupies so large a place to-day. The principal of the college was a first-rate man, the worthy Abbe X., who, not caring to dispense beans and bacon himself, had left the commissariat-department to a relative and had undertaken to teach the boys physics.

Let us attend one of his lessons. The subject is the barometer. The establishment happens to possess one, an old apparatus, covered with dust, hanging on the wall beyond the reach of profane hands and bearing on its face, in large letters, the words stormy, rain, fair.

'The barometer, ' says the good abbe, addressing his pupils, whom, in patriarchal fashion, he calls by their Christian names, 'the barometer tells us if the weather will be good or bad. You see the words written on the face—stormy, rain—do you see, Bastien? '

'Yes, I see, ' says Bastien, the most mischievous of the lot.

He has been looking through his book and knows more about the barometer than his teacher does.

'It consists, ' the abbe continues, 'of a bent glass tube filled with mercury, which rises and falls according to the weather. The shorter leg of this tube is open; the other... the other... well, we'll see. Here, Bastien, you're the tallest, get up on the chair and just feel with your finger if the long leg is open or closed. I can't remember for certain. '

Bastien climbs on the chair, stands as high as he can on tip-toe and fumbles with his finger at the top of the long column. Then, with a discreet smile spreading under the silky hairs of his dawning moustache:

'Yes, ' he says, 'that's it. The long leg is open at the top. There, I can feel the hole. '

And Bastien, to confirm his mendacious statement, keeps wriggling his forefinger at the top of the tube, while his fellow-conspirators suppress their enjoyment as best they can.

'That will do, ' says the unconscious abbe. 'You can get down, Bastien. Take a note of it, boys: the longer leg of the barometer is open; take a note of it. It's a thing you might forget; I had forgotten it myself. '

Thus was physics taught. Things improved, however: a master came and came to stay, one who knew that the long leg of the barometer is closed. I myself secured tables on which my pupils were able to write instead of scribbling on their knees; and, as my class was daily increasing in numbers, it ended by being divided into two. As soon as I had an assistant to look after the younger boys, things assumed a different aspect.

Among the subjects taught, one in particular appealed to both masters and pupils. This was open-air geometry, practical surveying. The college had none of the necessary outfit; but, with my fat pay — seven hundred francs a year, if you please! — I could not hesitate over the expense. A surveyor's chain and stakes, arrows, level, square and compass were bought with my money. A microscopic graphometer, not much larger than the palm of one's hand and costing perhaps five francs, was provided by the establishment. There was no tripod to it; and I had one made. In short, my equipment was complete.

And so, when May came, once every week we left the gloomy school-room for the fields. It was a regular holiday. The boys disputed for the honour of carrying the stakes, divided into bundles of three; and more than one shoulder, as we walked through the town, felt the reflected glory of those erudite rods. I myself—why conceal the fact? —was not without a certain satisfaction as I piously carried that most delicate and precious apparatus, the historic five-franc graphometer. The scene of operations was an untilled, flinty plain, a harmas, as we call it in the district. (Cf. "The Life of the Fly", by J. Henri Fabre, translated by Alexander Teixeira de Mattos: chapter 1. —Translator's Note.) Here, no curtain of green hedges or shrubs prevented me from keeping an eye upon my staff; here—an indispensable condition—I had not the irresistible temptation of the unripe apricots to fear for my scholars. The plain stretched far and wide, covered with nothing but flowering thyme and rounded pebbles. There was ample scope for every imaginable polygon; trapezes and triangles could be combined in all sorts of ways. The inaccessible distances had ample elbow-room; and there was even an old ruin, once a pigeon-house, that lent its perpendicular to the graphometer's performances.

Well, from the very first day, my attention was attracted by something suspicious. If I sent one of the boys to plant a stake, I would see him stop frequently on his way, bend down, stand up again, look about and stoop once more, neglecting his straight line and his signals. Another, who was told to pick up the arrows, would forget the iron pin and take up a pebble instead; and a third deaf to the measurements of angles, would crumble a clod of earth between his fingers. Most of them were caught licking a bit of straw. The polygon came to a full stop, the diagonals suffered. What could the mystery be?

I enquired; and everything was explained. A born searcher and observer, the scholar had long known what the master had not yet heard of, namely, that there was a big black Bee who made clay nests on the pebbles in the harmas. These nests contained honey; and my surveyors used to open them and empty the cells with a straw. The honey, although rather strong-flavoured, was most acceptable. I acquired a taste for it myself and joined the nest-hunters, putting off the polygon till later. It was thus that I first saw Reaumur's Mason-bee, knowing nothing of her history and nothing of her historian.

The magnificent Bee herself, with her dark-violet wings and black-velvet raiment, her rustic edifices on the sun-blistered pebbles amid the thyme, her honey, providing a diversion from the severities of the compass and the square, all made a great impression on my mind; and I wanted to know more than I had learnt from the schoolboys, which was just how to rob the cells of their honey with a straw. As it happened, my bookseller had a gorgeous work on insects for sale. It was called "Histoire naturelle des animaux articules", by de Castelnau (Francis Comte de Castelnau de la Porte (1812-1880), the naturalist and traveller. Castelnau was born in London and died at Melbourne. — Translator's Note.), E. Blanchard (Emile Blanchard (born 1820), author of various works on insects, Spiders, etc. —Translator's Note.) and Lucas (Pierre Hippolyte Lucas (born 1815), author of works on Moths and Butterflies, Crustaceans, etc. —Translator's Note.), and boasted a multitude of most attractive illustrations; but the price of it, the price of it! No matter: was not my splendid income supposed to cover everything, food for the mind as well as food for the body? Anything extra that I gave to the one I could save upon the other; a method of balancing painfully familiar to those who look to science for their livelihood. The purchase was effected. That day my professional emoluments were severely strained: I devoted a month's salary to the acquisition of the book. I had to resort to miracles of economy for some time to come before making up the enormous deficit.

The book was devoured; there is no other word for it. In it, I learnt the name of my black Bee; I read for the first time various details of the habits of insects; I found, surrounded in my eyes with a sort of halo, the revered names of Reaumur, Huber (Francois Huber (1750-1831), the Swiss naturalist, author of "Nouvelles observations sur les abeilles. " He early became blind from excessive study and conducted his scientific work thereafter with the aid of his wife. — Translator's Note.) and Leon Dufour (Jean Marie Leon Dufour (1780-1865), an army surgeon who served with distinction in several campaigns, and subsequently practised as a doctor in the Landes, where he attained great eminence as a naturalist. Fabre often refers to him as the Wizard of the Landes. Cf. "The Life of the Spider", by J. Henri Fabre, translated by Alexander Teixeira de Mattos: chapter 1; and "The Life of the Fly": chapter 1. —Translator's Note.); and, while I turned over the pages for the hundredth time, a voice within me seemed to whisper:

'You also shall be of their company! '

Ah, fond illusions, what has come of you? (The present essay is one of the earliest in the "Souvenirs Entomologiques. "—Translator's Note.)

But let us banish these recollections, at once sweet and sad, and speak of the doings of our black Bee. Chalicodoma, meaning a house of pebbles, concrete or mortar, would be a most satisfactory title, were it not that it has an odd sound to any one unfamiliar with Greek. The name is given to Bees who build their cells with materials similar to those which we employ for our own dwellings. The work of these insects is masonry; only it is turned out by a rustic mason more used to hard clay than to hewn stone. Reaumur, who knew nothing of scientific classification—a fact which makes many of his papers very difficult to understand—named the worker after her work and called our builders in dried clay Mason-bees, which describes them exactly.

We have two of them in our district: the Chalicodoma of the Walls (Chalicodoma muraria), whose history Reaumur gives us in a masterly fashion; and the Sicilian Chalicodoma (C. sicula) (For reasons that will become apparent after the reader has learnt their habits, the author also speaks of the Mason-bee of the Walls and the Sicilian Mason-bee as the Mason-bee of the Pebbles and the Mason-bee of the Sheds respectively. Cf. Chapter 4 footnote. —Translator's Note.), who is not peculiar to the land of Etna, as her name might suggest, but is also found in Greece, in Algeria and in the south of France, particularly in the department of Vaucluse, where she is one of the commonest Bees to be seen in the month of May. In the first species the two sexes are so unlike in colouring that a novice, surprised at observing them come out of the same nest, would at first take them for strangers to each other. The female is of a splendid velvety black, with dark-violet wings. In the male, the black velvet is replaced by a rather bright brick-red fleece. The second species, which is much smaller, does not show this contrast of colour: the two sexes wear the same costume, a general mixture of brown, red and grey, while the tips of the wings, washed with violet on a bronzed ground, recall, but only faintly, the rich purple of the first species. Both begin their labours at the same period, in the early part of May.

As Reaumur tells us, the Chalicodoma of the Walls in the northern provinces selects a wall directly facing the sun and one not covered with plaster, which might come off and imperil the future of the cells. She confides her buildings only to solid foundations, such as bare stones. I find her equally prudent in the south; but, for some reason which I do not know, she here generally prefers some other base to the stone of a wall. A rounded pebble, often hardly larger than one's fist, one of those cobbles with which the waters of the glacial period covered the terraces of the Rhone Valley, forms the most popular support. The extreme abundance of these sites might easily influence the Bee's choice: all our less elevated uplands, all our arid, thyme-clad grounds are nothing but water-worn stones cemented with red earth. In the valleys, the Chalicodoma has also the pebbles of the mountain-streams at her disposal. Near Orange, for instance, her favourite spots are the alluvia of the Aygues, with their carpets of smooth pebbles no longer visited by the waters. Lastly, if a cobble be wanting, the Mason-bee will establish her nest on any sort of stone, on a mile-stone or a boundary-wall.

The Sicilian Chalicodoma has an even greater variety of choice. Her most cherished site is the lower surface of the projecting tiles of a roof. There is not a cottage in the fields, however small, but shelters her nests under the eaves. Here, each spring, she settles in populous colonies, whose masonry, handed down from one generation to the next and enlarged year by year, ends by covering considerable surfaces. I have seen some of these nests, under the tiles of a shed, spreading over an area of five or six square yards. When the colony was hard at work, the busy, buzzing crowd was enough to make one giddy. The under side of a balcony also pleases the Mason-bee, as does the embrasure of a disused window, especially if it is closed by a blind whose slats allow her a free passage. But these are popular resorts, where hundreds and thousands of workers labour, each for herself. If she be alone, which happens pretty often, the Sicilian Mason-bee instals herself in the first little nook handy, provided that it supplies a solid foundation and warmth. As for the nature of this foundation, she does not seem to mind. I have seen her build on the bare stone, on bricks, on the wood of a shutter and even on the window-panes of a shed. One thing only does not suit her: the plaster of our houses. She is as prudent as her kinswoman and would fear the ruin of her cells, if she entrusted them to a support which might possibly fall.

Lastly, for reasons which I am still unable to explain to my own satisfaction, the Sicilian Mason-bee often changes the position of her building entirely, turning her heavy house of clay, which would seem to require the solid support of a rock, into an aerial dwelling. A hedge-shrub of any kind whatever—hawthorn, pomegranate, Christ's thorn—provides her with a foundation, usually as high as a man's head. The holm-oak and the elm give her a greater altitude. She chooses in the bushy clump a twig no thicker than a straw; and on this narrow base she constructs her edifice with the same mortar that she would employ under a balcony or the ledge of a roof. When finished, the nest is a ball of earth, bisected by the twig. It is the size of an apricot when the work of a single insect and of one's fist if several have collaborated; but this latter case is rare.

Both Bees use the same materials: calcareous clay, mingled with a little sand and kneaded into a paste with the mason's own saliva. Damp places, which would facilitate the quarrying and reduce the expenditure of saliva for mixing the mortar, are scorned by the Mason- bees, who refuse fresh earth for building even as our own builders refuse plaster and lime that have long lost their setting-properties. These materials, when soaked with pure moisture, would not hold properly. What is wanted is a dry dust, which greedily absorbs the disgorged saliva and forms with the latter's albuminous elements a sort of readily-hardening Roman cement, something in short resembling the cement which we obtain with quicklime and white of egg.

The mortar-quarry which the Sicilian Mason-bee prefers to work is a frequented highway, whose metal of chalky flints, crushed by the passing wheels, has become a smooth surface, like a continuous flagstone. Whether settling on a twig in a hedge or fixing her abode under the eaves of some rural dwelling, she always goes for her building-materials to the nearest path or road, without allowing herself to be distracted from her business by the constant traffic of people and cattle. You should see the active Bee at work when the road is dazzling white under the rays of a hot sun. Between the adjoining farm, which is the building-yard, and the road, in which the mortar is prepared, we hear the deep hum of the Bees perpetually crossing one another as they go to and fro. The air seems traversed by incessant trails of smoke, so straight and rapid is the worker's flight. Those on the way to the nest carry tiny pellets of mortar, the size of small shot; those who return at once settle on the driest and hardest spots. Their whole body aquiver, they scrape with the tips of their mandibles and rake with their front tarsi to extract atoms of earth and grains of sand, which, rolled between their teeth, become impregnated with saliva and form a solid mass. The work is pursued so vigorously that the worker lets herself be crushed under the feet of the passers-by rather than abandon her task.

On the other hand, the Mason-bee of the Walls, who seeks solitude, far from human habitations, rarely shows herself on the beaten paths, perhaps because these are too far from the places where she builds. So long as she can find dry earth, rich in small gravel, near the pebble chosen as the site of her nest, that is all she asks.

The Bee may either build an entirely new nest on a site as yet unoccupied, or she may use the cells of an old nest, after repairing them. Let us consider the former case first. After selecting her pebble, the Mason-bee of the Walls arrives with a little ball of mortar in her mandibles and lays it in a circular pad on the surface of the stone. The fore-legs and above all the mandibles, which are the mason's chief tools, work the material, which is kept plastic by the salivary fluid as this is gradually disgorged. In order to consolidate the clay, angular bits of gravel, the size of a lentil, are inserted separately, but only on the outside, in the as yet soft mass. This is the foundation of the structure. Fresh layers follow, until the cell has attained the desired height of two or three centimetres. (Three- quarters of an inch to one inch. —Translator's Note.)

Man's masonry is formed of stones laid one above the other and cemented together with lime. The Chalicodoma's work can bear comparison with ours. To economise labour and mortar, the Bee employs coarse materials, big pieces of gravel, which to her represent hewn stones. She chooses them carefully one by one, picks out the hardest bits, generally with corners which, fitting one into the other, give mutual support and contribute to the solidity of the whole. Layers of mortar, sparingly applied, hold them together. The outside of the cell thus assumes the appearance of a piece of rustic architecture, in which the stones project with their natural irregularities; but the inside, which requires a more even surface in order not to hurt the larva's tender skin, is covered with a coat of pure mortar. This inner whitewash, however, is put on without any attempt at art, indeed one might say that it is ladled on in great splashes; and the grub takes care, after finishing its mess of honey, to make itself a cocoon and hang the rude walls of its abode with silk. On the other hand, the Anthophorae and the Halicti, two species of Wild Bees whose grubs weave no cocoon, delicately glaze the inside of their earthen cells and give them the gloss of polished ivory.

The structure, whose axis is nearly always vertical and whose orifice faces upwards so as not to let the honey escape, varies a little in shape according to the supporting base. When set on a horizontal surface, it rises like a little oval tower; when fixed against an upright or slanting surface, it resembles the half of a thimble divided from top to bottom. In this case, the support itself, the pebble, completes the outer wall.

When the cell is finished, the Bee at once sets to work to victual it. The flowers round about, especially those of the yellow broom (Genista scoparia), which in May deck the pebbly borders of the mountain streams with gold, supply her with sugary liquid and pollen. She comes with her crop swollen with honey and her belly yellowed underneath with pollen dust. She dives head first into the cell; and for a few moments you see some spasmodic jerks which show that she is disgorging the honey-syrup. After emptying her crop, she comes out of the cell, only to go in again at once, but this time backwards. The Bee now brushes the lower side of her abdomen with her two hind-legs and rids herself of her load of pollen. Once more she comes out and once more goes in head first. It is a question of stirring the materials, with her mandibles for a spoon, and making the whole into a homogeneous mixture. This mixing-operation is not repeated after every journey: it takes place only at long intervals, when a considerable quantity of material has been accumulated.

The victualling is complete when the cell is half full. An egg must now be laid on the top of the paste and the house must be closed. All this is done without delay. The cover consists of a lid of pure mortar, which the Bee builds by degrees, working from the circumference to the centre. Two days at most appeared to me to be enough for everything, provided that no bad weather—rain or merely clouds—came to interrupt the labour. Then a second cell is built, backing on the first and provisioned in the same manner. A third, a fourth, and so on follow, each supplied with honey and an egg and closed before the foundations of the next are laid. Each task begun is continued until it is quite finished; the Bee never commences a new cell until the four processes needed for the construction of its predecessor are completed: the building, the victualling, the laying of the egg and the closing of the cell.

As the Mason-bee of the Walls always works by herself on the pebble which she has chosen and even shows herself very jealous of her site when her neighbours alight upon it, the number of cells set back to back upon one pebble is not large, usually varying between six and ten. Do some eight grubs represent the Bee's whole family? Or does she afterwards go and establish a more numerous progeny on other boulders? The surface of the same stone is spacious enough to provide a support for further cells if the number of eggs called for them; the Bee could build there very comfortably, without hunting for another site, without leaving the pebble to which she is attached by habit and long acquaintance. It seems to me therefore, exceedingly probable that the family is a small one and that it is all installed on the one stone, at any rate when the Mason-bee is building a new home.

The six to ten cells composing the cluster are certainly a solid dwelling, with their rustic gravel covering; but the thickness of their walls and lids, two millimetres (. 078 inch—Translator's Note.) at most, seems hardly sufficient to protect the grubs against the inclemencies of the weather. Set on its pebble in the open air, without any sort of shelter, the nest will have to undergo the heat of summer, which will turn each cell into a stifling furnace, followed by the autumn rains, which will slowly wear away the stonework, and by the winter frosts, which will crumble what the rains have respected. However hard the cement may be, can it possibly resist all these agents of destruction? And, even if it does resist, will not the grubs, sheltered by too thin a wall, have to suffer from excess of heat in summer and of cold in winter?

Without arguing all this out, the Bee nevertheless acts wisely. When all the cells are finished, she builds a thick cover over the group, formed of a material, impermeable to water and a bad conductor of heat, which acts as a protection at the same time against damp, heat and cold. This material is the usual mortar, made of earth mixed with saliva, but on this occasion with no small stones in it. The Bee applies it pellet by pellet, trowelful by trowelful, to the depth of a centimetre (. 39 inch — Translator's Note.) over the cluster of cells, which disappear entirely under the clay covering. When this is done, the nest has the shape of a rough dome, equal in size to half an orange. One would take it for a round lump of mud which had been thrown and half crushed against a stone and had then dried where it was. Nothing outside betrays the contents, no semblance of cells, no semblance of work. To the inexperienced eye, it is a chance splash of mud and nothing more.

This outer covering dries as quickly as do our hydraulic cements; and the nest is now almost as hard as a stone. It takes a knife with a strong blade to break open the edifice. And I would add, in conclusion, that, under its final form, the nest in no way recalls the original work, so much so that one would imagine the cells of the start, those elegant turrets covered with stucco-work, and the dome of the finish, looking like a mere lump of mud, to be the product of two different species. But scrape away the crust of cement and we shall easily recognize the cells below and their layers of tiny pebbles.

Instead of building a brand-new nest, on a hitherto unoccupied boulder, the Mason-bee of the Walls is always glad to make use of the old nests which have lasted through the year without suffering any damage worth mentioning. The mortar dome has remained very much what it was at the beginning, thanks to the solidity of the masonry, only it is perforated with a number of round holes, corresponding with the chambers, the cells inhabited by past generations of larvae. Dwellings such as these, which need only a little repair to put them in good condition, save a great deal of time and trouble; and the Mason-bees look out for them and do not decide to build new nests except when the old ones are wanting.

From one and the same dome there issue several inhabitants, brothers and sisters, ruddy males and black females, all the offspring of the same Bee. The males lead a careless existence, know nothing of work and do not return to the clay houses except for a brief moment to woo the ladies; nor do they reck of the deserted cabin. What they want is the nectar in the flower-cups, not mortar to mix between their mandibles. There remain the young mothers, who alone are charged with the future of the family. To which of them will the inheritance of the old nest revert? As sisters, they have equal rights to it: so our code would decide, since the day when it shook itself free of the old savage right of primogeniture. But the Mason-bees have not yet got beyond the primitive basis of property, the right of the first occupant.

When, therefore, the laying-time is at hand, the Bee takes possession of the first vacant nest that suits her and settles there; and woe to any sister or neighbour who shall henceforth dare to contest her ownership. Hot pursuits and fierce blows will soon put the newcomer to flight. Of the various cells that yawn like so many wells around the dome, only one is needed at the moment; but the Bee rightly calculates that the others will be useful presently for the other eggs; and she watches them all with jealous vigilance to drive away possible visitors. Indeed I do not remember ever seeing two Masons working on the same pebble.

The task is now very simple. The Bee examines the old cell to see what parts require repairing. She tears off the strips of cocoon hanging from the walls, removes the fragments of clay that fell from the ceiling when pierced by the last inhabitant to make her exit, gives a coat of mortar to the dilapidated parts, mends the opening a little; and that is all. Next come the storing, the laying of the eggs and the closing of the chamber. When all the cells, one after the other, are thus furnished, the outer cover, the mortar dome, receives a few repairs if it needs them; and the thing is done.

The Sicilian Mason-bee prefers company to a solitary life and establishes herself in her hundreds, very often in many thousands, under the tiles of a shed or the edge of a roof. These do not constitute a true society, with common interests to which all attend, but a mere gathering, where each works for herself and is not concerned with the rest, in short, a throng of workers recalling the swarm of a hive only by their numbers and their eagerness. The mortar employed is the same as that of the Mason-bee of the Walls, equally unyielding and waterproof, but thinner and without pebbles. The old nests are used first. Every free chamber is repaired, stocked and sealed up. But the old cells are far from sufficient for the population, which increases rapidly from year to year. Then, on the surface of the nest, whose chambers are hidden under the old general mortar covering, new cells are built, as the needs of the laying-time call for them. They are placed horizontally, or nearly so, side by side, with no attempt at orderly arrangement. Each architect has plenty of elbow-room and builds as and where she pleases, on the one condition that she does not hamper her neighbours' work; otherwise she can look out for rough handling from the parties interested. The cells, therefore, accumulate at random in this workyard where there is no organization. Their shape is that of a thimble divided down the middle; and their walls are completed either by the adjoining cells or by the surface of the old nest. Outside, they are rough and display successive layers of knotted cords corresponding with the different courses of mortar. Inside, the walls are flat without being smooth; later on, the grub's cocoon will make up for any lack of polish.

Each cell, as built, is stocked and walled up immediately, as we have seen with the Mason-bee of the Walls. This work goes on throughout the best part of May. All the eggs are laid at last; and then the Bees, without drawing distinctions between what does and what does not belong to them, set to work in common on a general protection for the colony. This is a thick coat of mortar, which fills up the gaps and covers all the cells. In the end, the common nest presents the appearance of a wide expanse of dry mud, with very irregular protuberances, thicker in the middle, the original nucleus of the establishment, thinner at the edges, where as yet there are only newly built cells, and varying greatly in dimensions according to the number of workers and therefore to the age of the nest first founded. Some of these nests are hardly larger than one's hand, while others occupy the greater part of the projecting edge of a roof and are measured by square yards.

When working alone, which is not unusual, on the shutter of a disused window, on a stone, or on a twig in some hedge, the Sicilian Chalicodoma behaves in just the same way. For instance, should she settle on a twig, the Bee begins by solidly cementing the base of her cell to the slight foundation. Next, the building rises, taking the form of a little upright turret. This first cell, when victualled and sealed, is followed by another, having as its support, in addition to the twig, the cells already built. From six to ten chambers are thus grouped side by side. Lastly, one coat of mortar covers everything, including the twig itself, which provides a firm mainstay for the whole.

CHAPTER 2
EXPERIMENTS.

As the nests of the Mason-bee of the Walls are erected on small-sized pebbles, which can be easily carried wherever you like and moved about from one place to another, without disturbing either the work of the builder or the repose of the occupants of the cells, they lend themselves readily to practical experiment, the only method that can throw a little light on the nature of instinct. To study the insect's mental faculties to any purpose, it is not enough for the observer to be able to profit by some happy combination of circumstances: he must know how to produce other combinations, vary them as much as possible and test them by substitution and interchange. Lastly, to provide science with a solid basis of facts, he must experiment. In this way, the evidence of formal records will one day dispel the fantastic legends with which our books are crowded: the Sacred Beetle (A Dung- beetle who rolls the manure of cattle into balls for his own consumption and that of his young. Cf. "Insect Life", by J. H. Fabre, translated by the author of "Mademoiselle Mori": chapters 1 and 2; and "The Life and Love of the Insect", by J. Henri Fabre, translated by Alexander Teixeira de Mattos: chapters 1 to 4. — Translator's Note.) calling on his comrades to lend a helping hand in dragging his pellet out of a rut; the Sphex (A species of Hunting Wasp. Cf. "Insect Life": chapters 6 to 12. —Translator's Note.) cutting up her Fly so as to be able to carry him despite the obstacle of the wind; and all the other fallacies which are the stock-in-trade of those who wish to see in the animal world what is not really there. In this way, again, materials will be prepared which will one day be worked up by the hand of a master and consign hasty and unfounded theories to oblivion.

Reaumur, as a rule, confines himself to stating facts as he sees them in the normal course of events and does not try to probe deeper into the insect's ingenuity by means of artificially produced conditions. In his time, everything had yet to be done; and the harvest was so great that the illustrious harvester went straight to what was most urgent, the gathering of the crop, and left his successors to examine the grain and the ear in detail. Nevertheless, in connection with the Chalicodoma of the Walls, he mentions an experiment made by his friend, Duhamel. (Henri Louis Duhamel du Monceau (1700-1781), a distinguished writer on botany and agriculture. —Translator's Note.) He tells us how a Mason-bee's nest was enclosed in a glass funnel, the mouth of which was covered merely with a bit of gauze. From it there issued three males, who, after vanquishing mortar as hard as stone, either never thought of piercing the flimsy gauze or else deemed the work beyond their strength. The three Bees died under the funnel. Reaumur adds that insects generally know only how to do what they have to do in the ordinary course of nature.

The experiment does not satisfy me, for two reasons: first, to ask workers equipped with tools for cutting clay as hard as granite to cut a piece of gauze does not strike me as a happy inspiration; you cannot expect a navvy's pick-axe to do the same work as a dressmaker's scissors. Secondly, the transparent glass prison seems to me ill- chosen. As soon as the insect has made a passage through the thickness of its earthen dome, it finds itself in broad daylight; and to it daylight means the final deliverance, means liberty. It strikes against an invisible obstacle, the glass; and to it glass is nothing at all and yet an obstruction. On the far side, it sees free space, bathed in sunshine. It wears itself out in efforts to fly there, unable to understand the futile nature of its attempts against that strange barrier which it cannot see. It perishes, at last, of exhaustion, without, in its obstinacy, giving a glance at the gauze closing the conical chimney. The experiment must be renewed under better conditions.

The obstacle which I select is ordinary brown paper, stout enough to keep the insect in the dark and thin enough not to offer serious resistance to the prisoner's efforts. As there is a great difference, in so far as the actual nature of the barrier is concerned, between a paper partition and a clay ceiling, let us begin by enquiring if the Mason-bee of the Walls knows how or rather is able to make her way through one of these partitions. The mandibles are pickaxes suitable for breaking through hard mortar: are they also scissors capable of cutting a thin membrane? This is the point to look into first of all.

In February, by which time the insect is in its perfect state, I take a certain number of cocoons, without damaging them, from their cells and insert them each in a separate stump of reed, closed at one end by the natural wall of the node and open at the other. These pieces of reed represent the cells of the nest. The cocoons are introduced with the insect's head turned towards the opening. Lastly, my artificial cells are closed in different ways. Some receive a stopper of kneaded clay, which, when dry, will correspond in thickness and consistency with the mortar ceiling of the natural nest. Others are plugged with a cylinder of sorghum, at least a centimetre (. 39 inch—Translator's Note.) thick; and the remainder with a disk of brown paper solidly fastened by the edge. All these bits of reed are placed side by side in a box, standing upright, with the roof of my making at the top. The insects, therefore, are in the exact position which they occupied in the nest. To open a passage, they must do what they would have done without my interference, they must break through the wall situated above their heads. I shelter the whole under a wide bell-glass and wait for the month of May, the period of the deliverance.

The results far exceed my anticipations. The clay stopper, the work of my fingers, is perforated with a round hole, differing in no wise from that which the Mason-bee contrives through her native mortar dome. The vegetable barrier, new to my prisoners, namely, the sorghum cylinder, also opens with a neat orifice, which might have been the work of a punch. Lastly, the brown-paper cover allows the Bee to make her exit not by bursting through, by making a violent rent, but once more by a clearly defined round hole. My Bees therefore are capable of a task for which they were not born; to come out of their reed cells they do what probably none of their race did before them; they perforate the wall of sorghum-pith, they make a hole in the paper barrier, just as they would have pierced their natural clay ceiling. When the moment comes to free themselves, the nature of the impediment does not stop them, provided that it be not beyond their strength; and henceforth the argument of incapacity cannot be raised when a mere paper barrier is in question.

In addition to the cells made out of bits of reed, I put under the bell-glass, at the same time, two nests which are intact and still resting on their pebbles. To one of them I have attached a sheet of brown paper pressed close against the mortar dome. In order to come out, the insect will have to pierce first the dome and then the paper, which follows without any intervening space. Over the other, I have placed a little brown paper cone, gummed to the pebble. There is here, therefore, as in the first case, a double wall—a clay partition and a paper partition—with this difference, that the two walls do not come immediately after each other, but are separated by an empty space of about a centimetre at the bottom, increasing as the cone rises.

The results of these two experiments are quite different. The Bees in the nest to which a sheet of paper was tightly stuck come out by piercing the two enclosures, of which the outer wall, the paper wrapper, is perforated with a very clean round hole, as we have already seen in the reed cells closed with a lid of the same material. We thus become aware, for the second time, that, when the Mason-bee is stopped by a paper barrier, the reason is not her incapacity to overcome the obstacle. On the other hand, the occupants of the nest covered with the cone, after making their way through the earthen dome, finding the sheet of paper at some distance, do not even try to perforate this obstacle, which they would have conquered so easily had it been fastened to the nest. They die under the cover without making any attempt to escape. Even so did Reaumur's Bees perish in the glass funnel, where their liberty depended only upon their cutting through a bit of gauze.

This fact strikes me as rich in inferences. What! Here are sturdy insects, to whom boring through granite is mere play, to whom a stopper of soft wood and a paper partition are walls quite easy to perforate despite the novelty of the material; and yet these vigorous housebreakers allow themselves to perish stupidly in the prison of a paper bag, which they could have torn open with one stroke of their mandibles! They are capable of tearing it, but they do not dream of doing so! There can be only one explanation of this suicidal inaction. The insect is well-endowed with tools and instinctive faculties for accomplishing the final act of its metamorphosis, namely, the act of emerging from the cocoon and from the cell. Its mandibles provide it with scissors, file, pick-axe and lever wherewith to cut, gnaw through and demolish either its cocoon and its mortar enclosure or any other not too obstinate barrier substituted for the natural covering of the nest. Moreover—and this is an important proviso, except for which the outfit would be useless—it has, I will not say the will to use those tools, but a secret stimulus inviting it to employ them. When the hour for the emergence arrives, this stimulus is aroused and the insect sets to work to bore a passage. It little cares in this case whether the material to be pierced be the natural mortar, sorghum-pith, or paper: the lid that holds it imprisoned does not resist for long. Nor even does it care if the obstacle be increased in thickness and a paper wall be added outside the wall of clay: the two barriers, with no interval between them, form but one to the Bee, who passes through them because the act of getting out is still one act and one only. With the paper cone, whose wall is a little way off, the conditions are changed, though the total thickness of wall is really the same. Once outside its earthen abode, the insect has done all that it was destined to do in order to release itself; to move freely on the mortar dome represents to it the end of the release, the end of the act of boring. Around the nest a new barrier appears, the wall made by the paper bag; but, in order to pierce this, the insect would have to repeat the act which it has just accomplished, the act which it is not intended to perform more than once in its life; it would, in short, have to make into a double act that which by nature is a single one; and the insect cannot do this, for the sole reason that it has not the wish to. The Mason-bee perishes for lack of the smallest gleam of intelligence. And this is the singular intellect in which it is the fashion nowadays to see a germ of human reason! The fashion will pass and the facts remain, bringing us back to the good old notions of the soul and its immortal destinies.

Reaumur tells us how his friend Duhamel, having seized a Mason-bee with a forceps when she had half entered the cell, head foremost, to fill it with pollen-paste, carried her to a closet at some distance from the spot where he captured her. The Bee got away from him in this closet and flew out through the window. Duhamel made straight for the nest. The Mason arrived almost as soon as he did and renewed her work. She only seemed a little wilder, says the narrator, in conclusion.

Why were you not here with me, revered master, on the banks of the Aygues, which is a vast expanse of pebbles for three-fourths of the year and a mighty torrent when it rains? I should have shown you something infinitely better than the fugitive escaping from the forceps. You would have witnessed—and in so doing, would have shared my surprise—not the brief flight of the Mason who, carried to the nearest room, releases herself and forthwith returns to her nest in that familiar neighbourhood, but long journeys through unknown country. You would have seen the Bee whom I carried to a great distance from her home, to quite unfamiliar ground, find her way back with a geographical sense of which the Swallow, the Martin and the Carrier-pigeon would not have been ashamed; and you would have asked yourself, as I did, what incomprehensible knowledge of the local map guides that mother seeking her nest.

To come to facts: it is a matter of repeating with the Mason-bee of the Walls my former experiments with the Cerceris-wasps (Cf. "Insect Life": chapter 19. —Translator's Note.), of carrying the insect, in the dark, a long way from its nest, marking it and then leaving it to its own resources. In case any one should wish to try the experiment for himself, I make him a present of my manner of operation, which may save him time at the outset. The insect intended for a long journey must obviously be handled with certain precautions. There must be no forceps employed, no pincers, which might maim a wing, strain it and weaken the power of flight. While the Bee is in her cell, absorbed in her work, I place a small glass test-tube over it. The Mason, when she flies away, rushes into the tube, which enables me, without touching her, to transfer her at once into a screw of paper. This I quickly close. A tin box, an ordinary botanizing-case, serves to convey the prisoners, each in her separate paper bag.

The most delicate business, that of marking each captive before setting her free, is left to be done on the spot selected for the starting-point. I use finely-powdered chalk, steeped in a strong solution of gum arabic. The mixture, applied to some part of the insect with a straw, leaves a white patch, which soon dries and adheres to the fleece. When a particular Mason-bee has to be marked so as to distinguish her from another in short experiments, such as I shall describe presently, I confine myself to touching the tip of the abdomen with my straw while the insect is half in the cell, head downwards. The slight touch is not noticed by the Bee, who continues her work quite undisturbed; but the mark is not very deep and moreover it is in a rather bad place for any prolonged experiment, for the Bee is constantly brushing her belly to detach the pollen and is sure to rub it off sooner or later. I therefore make another one, dropping the sticky chalk right in the middle of the thorax, between the wings.

It is hardly possible to wear gloves at this work: the fingers need all their deftness to take up the restless Bee delicately and to overpower her without rough pressure. It is easily seen that, though the job may yield no other profit, you are at least sure of being stung. The sting can be avoided with a little dexterity, but not always. You have to put up with it. In any case, the Mason-bee's sting is far less painful than that of the Hive-bee. The white spot is dropped on the thorax; the Mason flies off; and the mark dries on the journey.

I start with two Mason-bees of the Walls working at their nests on the pebbles in the alluvia of the Aygues, not far from Serignan. I carry them home with me to Orange, where I release them after marking them. According to the ordnance-survey map, the distance is about two and a half miles as the crow flies. The captives are set at liberty in the evening, at a time when the Bees begin to leave off work for the day. It is therefore probable that my two Bees will spend their night in the neighbourhood.

Next morning, I go to the nests. The weather is still too cool and the works are suspended. When the dew has gone, the Masons begin work. I see one, but without a white spot, bringing pollen to one of the nests which had been occupied by the travellers whom I am expecting. She is a stranger who, finding the cell whose owner I myself had exiled untenanted, has installed herself there and made it her property, not knowing that it is already the property of another. She has perhaps been victualling it since yesterday evening. Close upon ten o'clock, when the heat is at its full, the mistress of the house suddenly arrives: her title-deeds as the original occupant are inscribed for me in undeniable characters on her thorax white with chalk. Here is one of my travellers back.

Over waving corn, over fields all pink with sainfoin, she has covered the two miles and a half; and here she is, back at the nest, after foraging on the way, for the doughty creature arrives with her abdomen yellow with pollen. To come home again from the verge of the horizon is wonderful in itself; to come home with a well-filled pollen-brush is superlative economy. A journey, even a forced journey, always becomes a foraging-expedition.

She finds the stranger in the nest:

'What's this? I'll teach you! '

And the owner falls furiously upon the intruder, who possibly was meaning no harm. A hot chase in mid-air now takes place between the two Masons. From time to time, they hover almost without movement, face to face, with only a couple of inches separating them, and here, doubtless measuring forces with their eyes, they buzz insults at each other. Then they go back and alight on the nest in dispute, first one, then the other. I expect to see them come to blows, to make them draw their stings. But my hopes are disappointed: the duties of maternity speak in too imperious a voice for them to risk their lives and wipe out the insult in a mortal duel. The whole thing is confined to hostile demonstrations and a few insignificant cuffs.

Nevertheless, the real proprietress seems to derive double courage and double strength from the feeling that she is in her rights. She takes up a permanent position on the nest and receives the other, each time that she ventures to approach, with an angry quiver of her wings, an unmistakable sign of her righteous indignation. The stranger, at last discouraged, retires from the field. Forthwith the Mason resumes her work, as actively as though she had not just undergone the hardships of a long journey.

One more word on these quarrels about property. It is not unusual, when one Mason-bee is away on an expedition, for another, some homeless vagabond, to call at the nest, take a fancy to it and set to work on it, sometimes at the same cell, sometimes at the next, if there are several vacant, which is generally the case in the old nests. The first occupier, on her return, never fails to drive away the intruder, who always ends by being turned out, so keen and invincible is the mistress' sense of ownership. Reversing the savage Prussian maxim, 'Might is right, ' among the Mason-bees right is might, for there is no other explanation of the invariable retreat of the usurper, whose strength is not a whit inferior to that of the real owner. If she is less bold, this is because she has not the tremendous moral support of knowing herself in the right, which makes itself respected, among equals, even in the brute creation.

The second of my travellers does not reappear, either on the day when the first arrived or on the following days. I decide upon another experiment, on this occasion with five subjects. The starting-place is the same; and the place of arrival, the distance, the time of day, all remain unchanged. Of the five with whom I experiment, I find three at their nests next day; the two others are missing.

It is therefore fully established that the Mason-bee of the Walls, carried to a distance of two and a half miles and released at a place which she has certainly never seen before, is able to return to the nest. But why do first one out of two and then two out of five fail to join their fellows? What one can do cannot another do? Is there a difference in the faculty that guides them over unknown ground? Or is it not rather a difference in flying-power? I remember that my Bees did not all start off with the same vigour. Some were hardly out of my fingers before they darted furiously into the air, where I at once lost sight of them, whereas the others came dropping down a few yards away from me, after a short flight. The latter, it seems certain, must have suffered on the journey, perhaps from the heat concentrated in the furnace of my box. Or I may have hurt the articulation of the wings in marking them, an operation difficult to perform when you are guarding against stings. These are maimed, feeble creatures, who will linger in the sainfoin-fields close by, and not the powerful aviators required by the journey.

The experiment must be tried again, taking count only of the Bees who start off straight from between my fingers with a clean, vigorous flight. The waverers, the laggards who stop almost at once on some bush shall be left out of the reckoning. Moreover, I will do my best to estimate the time taken in returning to the nest. For an experiment of this kind, I need plenty of subjects, as the weak and the maimed, of whom there may be many, are to be disregarded. The Mason-bee of the Walls is unable to supply me with the requisite number: there are not enough of her; and I am anxious not to interfere too much with the little Aygues-side colony, for whom I have other experiments in view. Fortunately, I have at my own place, under the eaves of a shed, a magnificent nest of Chalicodoma sicula in full activity. I can draw to whatever extent I please on the populous city. The insect is small, less than half the size of C. muraria, but no matter: it will deserve all the more credit if it can traverse the two miles and a half in store for it and find its way back to the nest. I take forty Bees, isolating them, as usual, in screws of paper.

In order to reach the nest, I place a ladder against the wall: it will be used by my daughter Aglae and will enable her to mark the exact moment of the return of the first Bee. I set the clock on the mantelpiece and my watch at the same time, so that we may compare the instant of departure and of arrival. Things being thus arranged, I carry off my forty captives and go to the identical spot where C. muraria works, in the pebbly bed of the Aygues. The trip will have a double object: to observe Reaumur's Mason and to set the Sicilian Mason at liberty. The latter, therefore, will also have two and a half miles to travel home.

At last my prisoners are released, all of them being first marked with a big white dot in the middle of the thorax.

You do not come off scot-free when handling one after the other forty wrathful Bees, who promptly unsheathe and brandish their poisoned stings. The stab is but too often given before the mark is made. My smarting fingers make movements of self-defence which my will is not always able to control. I take hold with greater precaution for myself than for the insect; I sometimes squeeze harder than I ought to if I am to spare my travellers. To experiment so as to lift, if possible, a tiny corner of the veil of truth is a fine and noble thing, a mighty stimulant in the face of danger; but still one may be excused for displaying some impatience when it is a matter of receiving forty stings in one's fingers at one short sitting. If any man should reproach me for being too careless with my thumbs, I would suggest that he should have a try: he can then judge for himself the pleasures of the situation.

To cut a long story short, either through the fatigue of the journey, or through my fingers pressing too hard and perhaps injuring some articulations, only twenty out of my forty Bees start with a bold, vigorous flight. The others, unable to keep their balance, wander about on the nearest bit of grass or remain on the osier-shoots on which I have placed them, refusing to fly even when I tickle them with a straw. These weaklings, these cripples, these incapables injured by my fingers must be struck off my list. Those who started with an unhesitating flight number about twenty. That is ample.

At the actual moment of departure, there is nothing definite about the direction taken, none of that straight flight to the nest which the Cerceris-wasps once showed me in similar circumstances. As soon as they are liberated, the Mason-bees flee as though scared, some in one direction, some in exactly the opposite direction. Nevertheless, as far as their impetuous flight allows, I seem to perceive a quick return on the part of those Bees who have started flying towards a point opposite to their home; and the majority appear to me to be making for those blue distances where their nest lies. I leave this question with certain doubts which are inevitable in the case of insects which I cannot follow with my eyes for more than twenty yards.

Hitherto, the operation has been favoured by calm weather; but now things become complicated. The heat is stifling and the sky becomes stormy. A stiff breeze springs up, blowing from the south, the very direction which my Bees must take to return to the nest. Can they overcome this opposing current and cleave the aerial torrent with their wings? If they try, they will have to fly close to the ground, as I now see the Bees do who continue their foraging; but soaring to lofty regions, whence they can obtain a clear view of the country, is, so it seems to me, prohibited. I am therefore very apprehensive as to the success of my experiment when I return to Orange, after first trying to steal some fresh secret from the Aygues Mason-bee of the Pebbles.

I have scarcely reached the house before Aglae greets me, her cheeks flushed with excitement:

'Two! ' she cries. 'Two came back at twenty minutes to three, with a load of pollen under their bellies! '

A friend of mine had appeared upon the scene, a grave man of the law, who on hearing what was happening, had neglected code and stamped paper and insisted upon also being present at the arrival of my Carrier-pigeons. The result interested him more than his case about a party-wall. Under a tropical sun, in a furnace heat reflected from the wall of the shed, every five minutes he climbed the ladder bare- headed, with no other protection against sunstroke than his thatch of thick, grey locks. Instead of the one observer whom I had posted, I found two good pairs of eyes watching the Bees' return.

I had released my insects at about two o'clock; and the first arrivals returned to the nest at twenty minutes to three. They had therefore taken less than three-quarters of an hour to cover the two miles and a half, a very striking result, especially when we remember that the Bees did some foraging on the road, as was proved by the yellow pollen on their bellies, and that, on the other hand, the travellers' flight must have been hindered by the wind blowing against them. Three more came home before my eyes, each with her load of pollen, an outward and visible sign of the work done on the journey. As it was growing late, our observations had to cease. When the sun goes down, the Mason-bees leave the nest and take refuge somewhere or other, perhaps under the tiles of the roofs, or in little corners of the walls. I could not reckon on the arrival of the others before work was resumed, in the full sunshine.

Next day, when the sun recalled the scattered workers to the nest, I took a fresh census of Bees with a white spot on the thorax. My success exceeded all my hopes: I counted fifteen, fifteen of the transported prisoners of the day before, storing their cells or building as though nothing out of the way had happened. The weather had become more and more threatening; and now the storm burst and was followed by a succession of rainy days which prevented me from continuing.

The experiment suffices as it stands. Of some twenty Bees who had seemed fit to make the long journey when I released them, fifteen at least had returned: two within the first hour, three in the course of the evening and the rest next morning. They had returned in spite of having the wind against them and—a graver difficulty still—in spite of being unacquainted with the locality to which I had transported them. There is, in fact, no doubt that they were setting eyes for the first time on those osier-beds of the Aygues which I had selected as the starting-point. Never would they have travelled so far afield of their own accord, for everything that they want for building and victualling under the roof of my shed is within easy reach. The path at the foot of the wall supplies the mortar; the flowery meadows surrounding my house furnish nectar and pollen. Economical of their time as they are, they do not go flying two miles and a half in search of what abounds at a few yards from the nest. Besides, I see them daily taking their building-materials from the path and gathering their harvest on the wild-flowers, especially on the meadow sage. To all appearance, their expeditions do not cover more than a radius of a hundred yards or so. Then how did my exiles return? What guided them? It was certainly not memory, but some special faculty which we must content ourselves with recognizing by its astonishing effects without pretending to explain it, so greatly does it transcend our own psychology.

CHAPTER 3
EXCHANGING THE NESTS.

Let us continue our series of tests with the Mason-bee of the Walls. Thanks to its position on a pebble which we can move at will, the nest of this Bee lends itself to most interesting experiments. Here is the first: I shift a nest from its place, that is to say, I carry the pebble which serves as its support to a spot two yards away. As the edifice and its base form but one, the removal is performed without the smallest disturbance of the cells. I lay the boulder in an exposed place where it is well in view, as it was on its original site. The Bee returning from her harvest cannot fail to see it.

In a few minutes, the owner arrives and goes straight to where the nest stood. She hovers gracefully over the vacant site, examines and alights upon the exact spot where the stone used to lie. Here she walks about for a long time, making persistent searches; then the Bee takes wing and flies away to some distance. Her absence is of short duration. Here she is back again. The search is resumed, walking and flying, and always on the site which the nest occupied at first. A fresh fit of exasperation, that is to say, an abrupt flight across the osier-bed, is followed by a fresh return and a renewal of the vain search, always upon the mark left by the shifted pebble. These sudden departures, these prompt returns, these persevering inspections of the deserted spot continue for a long time, a very long time, before the Mason is convinced that her nest is gone. She has certainly seen it, has seen it over and over again in its new position, for sometimes she has flown only a few inches above it; but she takes no notice of it. To her, it is not her nest, but the property of another Bee.

Often the experiment ends without so much as a single visit to the boulder which I have moved two or three yards away: the Bee goes off and does not return. If the distance be less, a yard for instance, the Mason sooner or later alights on the stone which supports her abode. She inspects the cell which she was building or provisioning a little while before, repeatedly dips her head into it, examines the surface of the pebble step by step and, after long hesitations, goes and resumes her search on the site where the home ought to be. The nest that is no longer in its natural place is definitely abandoned, even though it be but a yard away from the original spot. Vainly does the Bee settle on it time after time: she cannot recognize it as hers. I was convinced of this on finding it, several days after the experiment, in just the same condition as when I moved it. The open cell half-filled with honey was still open and was surrendering its contents to the pillaging Ants; the cell that was building had remained unfinished, with not a single layer added to it. The Bee, obviously, may have returned to it; but she had not resumed work upon it. The transplanted dwelling was abandoned for good and all.

I will not deduce the strange paradox that the Mason-bee, though capable of finding her nest from the verge of the horizon, is incapable of finding it at a yard's distance: I interpret the occurrence as meaning something quite different. The proper inference appears to me to be this: the Bee retains a rooted impression of the site occupied by the nest and returns to it with unwearying persistence even when the nest is gone. But she has only a very vague notion of the nest itself. She does not recognize the masonry which she herself has erected and kneaded with her saliva; she does not know the pollen-paste which she herself has stored. In vain she inspects her cell, her own handiwork; she abandons it, refusing to acknowledge it as hers, once the spot whereon the pebble rests is changed.

Insect memory, it must be confessed, is a strange one, displaying such lucidity in its general acquaintance with locality and such limitations in its knowledge of the dwelling. I feel inclined to call it topographical instinct: it grasps the map of the country and not the beloved nest, the home itself. The Bembex-wasps (Cf. "Insect Life": chapters 16 to 19. —Translator's Note.) have already led us to a like conclusion. When the nest is laid open, these Wasps become wholly indifferent to the family, to the grub writhing in agony in the sun. They do not recognize it. What they do recognize, what they seek and find with marvellous precision, is the site of the entrance-door of which nothing at all is left, not even the threshold.

If any doubts remained as to the incapacity of the Mason-bee of the Walls to know her nest other than by the place which the pebble occupies on the ground, here is something to remove them: for the nest of one Mason-bee, I substitute that of another, resembling it as closely as possible in respect to both masonry and storage. This exchange and those of which I shall speak presently are of course made in the owner's absence. The Bee settles without hesitation in this nest which is not hers, but which stands where the other did. If she was building, I offer her a cell in process of building. She continues the masonry with the same care and the same zeal as if the work already done were her own work. If she was fetching honey and pollen, I offer her a partly-provisioned cell. She continues her journeys, with honey in her crop and pollen under her belly, to finish filling another's warehouse. The Bee, therefore, does not suspect the exchange; she does not distinguish between what is her property and what is not; she imagines that she is still working at the cell which is really hers.

After leaving her for a time in possession of the strange nest, I give her back her own. This fresh change passes unperceived by the Bee: the work is continued in the cell restored to her at the point which it had reached in the substituted cell. I once more replace it by the strange nest; and again the insect persists in continuing its labour. By thus constantly interchanging the strange nest and the proper nest, without altering the actual site, I thoroughly convinced myself of the Bee's inability to discriminate between what is her work and what is not. Whether the cell belong to her or to another, she labours at it with equal zest, so long as the basis of the edifice, the pebble, continues to occupy its original position.

The experiment receives an added interest if we employ two neighbouring nests the work on which is about equally advanced. I move each to where the other stood. They are not much more than thirty inches a part. In spite of their being so near to each other that it is quite possible for the insects to see both homes at once and choose between them, each Bee, on arriving, settles immediately on the substituted nest and continues her work there. Change the two nests as often as you please and you shall see the two Mason-bees keep to the site which they selected and labour in turn now at their own cell and now at the other's.

One might think that the cause of this confusion lies in a close resemblance between the two nests, for at the start, little expecting the results which I was to obtain, I used to choose the nests which I interchanged as much alike as possible, for fear of disheartening the Bees. I need not have taken this precaution: I was giving the insect credit for a perspicacity which it does not possess. Indeed, I now take two nests which are extremely unlike each other, the only point of resemblance being that, in each case, the toiler finds a cell in which she can continue the work which she is actually doing. The first is an old nest whose dome is perforated with eight holes, the apertures of the cells of the previous generation. One of these cells has been repaired; and the Bee is busy storing it. The second is a nest of recent construction, which has not received its mortar dome and consists of a single cell with its stucco covering. Here too the insect is busy hoarding pollen-paste. No two nests could present greater differences: one with its eight empty chambers and its spreading clay dome; the other with its single bare cell, at most the size of an acorn.

Well, the two Mason-bees do not hesitate long in front of these exchanged nests, not three feet away from each other. Each makes for the site of her late home. One, the original owner of the old nest, finds nothing but a solitary cell. She rapidly inspects the pebble and, without further formalities, first plunges her head into the strange cell, to disgorge honey, and then her abdomen, to deposit pollen. And this is not an action due to the imperative need of ridding herself as quickly as possible, no matter where, of an irksome load, for the Bee flies off and soon comes back again with a fresh supply of provender, which she stores away carefully. This carrying of provisions to another's larder is repeated as often as I permit it. The other Bee, finding instead of her one cell a roomy structure consisting of eight apartments, is at first not a little embarrassed. Which of the eight cells is the right one? In which is the heap of paste on which she had begun? The Bee therefore visits the chambers one by one, dives right down to the bottom and ends by finding what she seeks, that is to say, what was in her nest when she started on her last journey, the nucleus of a store of food. Thenceforward she behaves like her neighbour and goes on carrying honey and pollen to the warehouse which is not of her constructing.

Restore the nests to their original places, exchange them yet once again and both Bees, after a short hesitation which the great difference between the two nests is enough to explain, will pursue the work in the cell of her own making and in the strange cell alternately. At last the egg is laid and the sanctuary closed, no matter what nest happens to be occupied at the moment when the provisioning reaches completion. These incidents are sufficient to show why I hesitate to give the name of memory to the singular faculty that brings the insect back to her nest with such unerring precision and yet does not allow her to distinguish her work from some one else's, however great the difference may be.

We will now experiment with Chalicodoma muraria from another psychological point of view. Here is a Mason-bee building; she is at work on the first course of her cell. I give her in exchange a cell not only finished as a structure, but also filled nearly to the top with honey. I have just stolen it from its owner, who would not have been long before laying her egg in it. What will the Mason do in the presence of this munificent gift, which saves her the trouble of building and harvesting? She will leave the mortar no doubt, finish storing the Bee-bread, lay her egg and seal up. A mistake, an utter mistake: our logic is not the logic of the insect, which obeys an inevitable, unconscious prompting. It has no choice as to what it shall do; it cannot discriminate between what is and what is not advisable; it glides, as it were, down an irresistible slope prepared beforehand to bring it to a definite end. This is what the facts that still remain to be stated proclaim with no uncertain voice.

The Bee who was building and to whom I offer a cell ready-built and full of honey does not lay aside her mortar for that. She was doing mason's work; and, once on that tack, guided by the unconscious impulse, she has to keep masoning, even though her labour be useless, superfluous and opposed to her interests. The cell which I give her is certainly perfect, looked upon as a building, in the opinion of the master-builder herself, since the Bee from whom I took it was completing the provision of honey. To touch it up, especially to add to it, is useless and, what is more, absurd. No matter: the Bee who was masoning will mason. On the aperture of the honey-store she lays a first course of mortar, followed by another and yet another, until at last the cell is a third taller then the regulation height. The masonry-task is now done, not as perfectly, it is true, as if the Bee had gone on with the cell whose foundations she was laying at the moment when I exchanged the nests, but still to an extent which is more than enough to prove the overpowering impulse which the builder obeys. Next comes the victualling, which is also cut short, lest the honey-store swelled by the joint contributions of the two Bees should overflow. Thus the Mason-bee who is beginning to build and to whom we give a complete cell, a cell filled with honey, makes no change in the order of her work: she builds first and then victuals. Only she shortens her work, her instinct warning her that the height of the cell and the quantity of honey are beginning to assume extravagant proportions.

The converse is equally conclusive. To a Mason-bee engaged in victualling I give a nest with a cell only just begun and not at all fit to receive the paste. This cell, with its last course still wet with its builder's saliva, may or may not be accompanied by other cells recently closed up, each with its honey and its egg. The Bee, finding this in the place of her half-filled honey-store, is greatly perplexed what to do when she comes with her harvest to this unfinished, shallow cup, in which there is no place to put the honey. She inspects it, measures it with her eyes, tries it with her antennae and recognizes its insufficient capacity. She hesitates for a long time, goes away, comes back, flies away again and soon returns, eager to deposit her treasure. The insect's embarrassment is most evident; and I cannot help saying, inwardly:

'Get some mortar, get some mortar and finish making the warehouse. It will only take you a few moments; and you will have a cupboard of the right depth. '

The Bee thinks differently: she was storing her cell and she must go on storing, come what may. Never will she bring herself to lay aside the pollen-brush for the trowel; never will she suspend the foraging which is occupying her at this moment to begin the work of construction which is not yet due. She will rather go in search of a strange cell, in the desired condition, and slip in there to deposit her honey, at the risk of meeting with a warm reception from the irate owner. She goes off, in fact, to try her luck. I wish her success, being myself the cause of this desperate act. My curiosity has turned an honest worker into a robber.

Things may take a still more serious turn, so invincible, so imperious is the desire to have the booty stored in a safe place without delay. The uncompleted cell which the Bee refuses to accept instead of her own finished warehouse, half-filled with honey, is often, as I said, accompanied by other cells, not long closed, each containing its Bee-bread and its egg. In this case, I have sometimes, though not always, witnessed the following: when once the Bee realises the shortcomings of the unfinished nest, she begins to gnaw the clay lid closing one of the adjoining cells. She softens a part of the mortar cover with saliva and patiently, atom by atom, digs through the hard wall. It is very slow work. A good half-hour elapses before the tiny cavity is large enough to admit a pin's head. I wait longer still. Then I lose patience; and, fully convinced that the Bee is trying to open the store-room, I decide to help her to shorten the work. The upper part of the cell comes away with it, leaving the edges badly broken. In my awkwardness, I have turned an elegant vase into a wretched cracked pot.

I was right in my conjecture: the Bee's intention was to break open the door. Straight away, without heeding the raggedness of the orifice, she settles down in the cell which I have opened for her. Time after time, she fetches honey and pollen, though the larder is already fully stocked. Lastly, she lays her egg in this cell which already contains an egg that is not hers, having done which she closes the broken aperture to the best of her ability. So this purveyor had neither the knowledge nor the power to bow to the inevitable. I had made it impossible for her to go on with her purveying, unless she first completed the unfinished cell substituted for her own. But she did not retreat before that impossible task. She accomplished her work, but in the absurdest way: by injuriously trespassing upon another's property, by continuing to store provisions in a cupboard already full to overflowing, by laying her egg in a cell in which the real owner had already laid and lastly by hurriedly closing an orifice that called for serious repairs. What better proof could be wished of the irresistible propensity which the insect obeys?

Lastly, there are certain swift and consecutive actions so closely interlinked that the performance of the second demands a previous repetition of the first, even when this action has become useless. I have already described how the Yellow-winged Sphex (Cf. "Insect Life": chapters 6 to 9. — Translator's Note.) persists in descending into her burrow alone, after depositing at its edge the Cricket whom I maliciously at once remove. Her repeated discomfitures do not make her abandon the preliminary inspection of the home, an inspection which becomes quite useless when renewed for the tenth or twentieth time. The Mason-bee of the Walls shows us, under another form, a similar repetition of an act which is useless in itself, but which is the compulsory preface to the act that follows. When arriving with her provisions, the Bee performs a twofold operation of storing. First, she dives head foremost into the cell, to disgorge the contents of her crop; next, she comes out and at once goes in again backwards, to brush her abdomen and rub off the load of pollen. At the moment when the insect is about to enter the cell tail first, I push her aside gently with a straw. The second act is thus prevented. The Bee now begins the whole performance over again, that is to say, she once more dives head first to the bottom of the cell, though she has nothing left to disgorge, as her crop has just been emptied. When this is done, it is the belly's turn. I instantly push her aside again. The insect repeats its proceedings, still entering head first; I also repeat my touch of the straw. And this can go on as long as the observer pleases. Pushed aside at the moment when she is about to insert her abdomen into the cell, the Bee goes back to the opening and persists in going down head first to begin with. Sometimes, she descends to the bottom, sometimes only half-way, sometimes again she only pretends to descend, just bending her head into the aperture; but, whether completed or not, this action, for which there is no longer any motive, since the honey has already been disgorged, invariably precedes the entrance backwards to deposit the pollen. It is almost the movement of a machine whose works are only set going when the driving-wheel begins to revolve.

CHAPTER 4
MORE ENQUIRIES INTO MASON-BEES.

This chapter was to have taken the form of a letter addressed to Charles Darwin, the illustrious naturalist who now lies buried beside Newton in Westminster Abbey. It was my task to report to him the result of some experiments which he had suggested to me in the course of our correspondence: a very pleasant task, for, though facts, as I see them, disincline me to accept his theories, I have none the less the deepest veneration for his noble character and his scientific honesty. I was drafting my letter when the sad news reached me: Darwin was dead; after searching the mighty question of origins, he was now grappling with the last and darkest problem of the hereafter. (Darwin died at Down, in Kent, on the 19th of April 1882. —Translator's Note.) I therefore abandon the epistolary form, which would be unwarranted in view of that grave at Westminster. A free and impersonal statement shall set forth what I intended to relate in a more academic manner.

One thing, above all, had struck the English scientist on reading the first volume of my "Souvenirs entomologiques", namely, the Mason-bees' faculty of knowing the way back to their nests after being carried to great distances from home. What sort of compass do they employ on their return journeys? What sense guides them? The profound observer thereupon spoke of an experiment which he had always longed to make with Pigeons and which he had always neglected making, absorbed as he was by other interests. This experiment, he thought, I might attempt with my Bees. Substitute the insect for the bird; and the problem remained the same. I quote from his letter the passage referring to the trial which he wished made:

'Allow me to make a suggestion in relation to your wonderful account of insects finding their way home. I formerly wished to try it with pigeons; namely, to carry the insects in their paper cornets about a hundred paces in the opposite direction to that which you intended ultimately to carry them, but before turning round to return, to put the insects in a circular box with an axle which could be made to revolve very rapidly first in one direction and then in another, so as to destroy for a time all sense of direction in the insects. I have sometimes imagined that animals may feel in which direction they were at the first start carried. '

This method of experimenting seemed to me very ingeniously conceived. Before going west, I walk eastwards. In the darkness of their paper bags, the mere fact that I am moving them gives my prisoners a sense of the direction in which I am taking them. If nothing happened to disturb this first impression, the insect would be guided by it in returning. This would explain the homing of my Mason-bees carried to a distance of two or three miles amid strange surroundings. But, when the insects have been sufficiently impressed by their conveyance to the east, there comes the rapid twirl, first this way round, then that. Bewildered by all these revolutions first in one direction and then in another, the insect does not know that I have turned round and remains under its original impression. I am now taking it to the west, when it believes itself to be still travelling towards the east. Under the influence of this impression; the insect is bound to lose its bearings. When set free, it will fly in the opposite direction to its home, which it will never find again.

This result seemed to me the more probable inasmuch as the statements of the country-folk around me were all of a nature to confirm my hopes. Favier (The author's gardener and factotum. Cf. "The Life of the Fly": chapter 4. —Translator's Note.), the very man for this sort of information, was the first to put me on the track. He told me that, when people want to move a Cat from one farm to another at some distance, they place the animal in a bag which they twirl rapidly at the moment of starting, thus preventing the animal from returning to the house which it has quitted. Many others, besides Favier, described the same practice to me. According to them, this twirling round in a bag was an infallible expedient: the bewildered Cat never returned. I communicated what I had learnt to England, I wrote to the sage of Down and told him how the peasant had anticipated the researches of science. Charles Darwin was amazed; so was I; and we both of us almost reckoned on a success.

These preliminaries took place in the winter; I had plenty of time to prepare for the experiment which was to be made in the following May.

'Favier, ' I said, one day, to my assistant, 'I shall want some of those nests. Go and ask our next-door neighbour's leave and climb to the roof of his shed, with some new tiles and some mortar, which you can fetch from the builder's. Take a dozen tiles from the roof, those with the biggest nests on them, and put the new ones in their place. '

Things were done accordingly. My neighbour assented with a good grace to the exchange of tiles, for he himself is obliged, from time to time, to demolish the work of the Mason-bee, unless he would risk seeing his roof fall in sooner or later. I was merely forestalling a repair which became more urgent every year. That same evening, I was in possession of twelve magnificent rectangular blocks of nest, each lying on the convex surface of a tile, that is to say, on the surface looking towards the inside of the shed. I had the curiosity to weigh the largest: it turned the scale at thirty-five pounds. Now the roof whence it came was covered with similar masses, adjoining one another, over a stretch of some seventy tiles. Reckoning only half the weight, so as to strike an average between the largest and the smallest lumps, we find the total weight of the Bee's masonry to amount to three- quarters of a ton. And, even so, people tell me that they have seen this beaten elsewhere. Leave the Mason-bee to her own devices, in the spot that suits her; allow the work of many generations to accumulate; and, one fine day, the roof will break down under the extra burden. Let the nests grow old; let them fall to pieces when the damp gets into them; and you will have chunks tumbling on your head big enough to crack your skull. There you see the work of a very little-known insect. (The insect is so little known that I made a serious mistake when treating of it in the first volume of these "Souvenirs. " Under my erroneous denomination of Chalicodoma sicula are really comprised two species, one building its nests in our dwellings and particularly under the tiles of outhouses, the other building its nests on the branches of shrubs. The first species has received various names, which are, in order of priority: Chalicodoma pyrenaica, LEP. (Megachile); Chalicodoma pyrrhopeza, GERSTACKER; Chalicodoma rufitarsis, GIRAUD. It is a pity that the name occupying the first place should lend itself to misconception. I hesitate to apply the epithet of Pyrenean to an insect which is much less common in the Pyrenees than in my own district. I shall call it the Chalicodoma, or Mason-bee, of the Sheds. There is no objection to the use of this name in a book where the reader prefers lucidity to the tyranny of systematic entomology. The second species, that which builds its nests on the branches, is Chalicodoma rufescens, J. PEREZ. For a like reason, I shall call it the Chalicodoma of the Shrubs. I owe these corrections to the kindness of Professor Jean Perez, of Bordeaux, who is so well-versed in the lore of Wasps and Bees. —Author's Note.)

These treasures were insufficient, not in regard to quantity, but in regard to quality, for the main object which I had in view. They came from the nearest house, separated from mine by a little field planted with corn and olive-trees. I had reason to fear that the insects issuing from those nests might be hereditarily influenced by their ancestors, who had lived in the shed for many a long year. The Bee, when carried to a distance, would perhaps come back, guided by the inveterate family habit; she would find the shed of her lineal predecessors and thence, without difficulty, reach her nest. As it is the fashion nowadays to assign a prominent part to these hereditary influences, I must eliminate them from my experiments. I want strange Bees, brought from afar, whose return to the place of their birth can in no way assist their return to the nest transplanted to another site.

Favier took the business in hand. He had discovered on the banks of the Aygues, at some miles from the village, a deserted hut where the Mason-bees had established themselves in a numerous colony. He proposed to take the wheelbarrow, in which to move the blocks of cells; but I objected: the jolting of the vehicle over the rough paths might jeopardise the contents of the cells. A basket carried on the shoulder was deemed safer. Favier took a man to help him and set out. The expedition provided me with four well-stocked tiles. It was all that the two men were able to carry between them; and even then I had to stand treat on their arrival: they were utterly exhausted. Le Vaillant tells us of a nest of Republicans (Social Weaver-birds. — Translator's Note.) with which he loaded a wagon drawn by two oxen. My Mason-bee vies with the South-African bird: a yoke of Oxen would not have been too many to move the whole of that nest from the banks of the Aygues.

The next thing is to place my tiles. I want to have them under my eyes, in a position where I can watch them easily and save myself the worries of earlier days: going up and down ladders, standing for hours at a stretch on a narrow rung that hurt the soles of my feet and risking sunstroke up against a scorching wall. Moreover, it is necessary that my guests should feel almost as much at home with me as where they come from. I must make life pleasant for them, if I should have them grow attached to the new dwelling. And I happen to have the very thing for them.

Under the leads of my house is a wide arch, the sides of which get the sun, while the back remains in the shade. There is something for everybody: the shade for me, the sunlight for my boarders. We fasten a stout hook to each tile and hang it on the wall, on a level with our eyes. Half my nests are on the right, half on the left. The general effect is rather original. Any one walking in and seeing my show for the first time begins by taking it for a display of smoked provisions, gammons of some outlandish bacon curing in the sun. On perceiving his mistake, he falls into raptures at these new hives of mine. The news spreads through the village and more than one pokes fun at it. They look upon me as a keeper of hybrid Bees:

'I wonder what he's going to make out of that! ' say they.

My hives are in full swing before the end of April. When the work is at its height, the swarm becomes a little eddying, buzzing cloud. The arch is a much-frequented passage: it leads to a store-room for various household provisions. The members of my family bully me at first for establishing this dangerous commonwealth within the precincts of our home. They dare not go to fetch things: they would have to pass through a swarm of Bees; and then... look out for stings! There is nothing for it but to prove, once and for all, that the danger does not exist, that mine is a most peaceable Bee, incapable of stinging so long as she is not startled. I bring my face close to one of the clay nests, so as almost to touch it, while it is black with Masons at work; I let my fingers wander through the ranks, I put a few Bees on my hand, I stand in the thick of the whirling crowd and never a prick do I receive. I have long known their peaceful character. Time was when I used to share the common fears, when I hesitated before venturing into a swarm of Anthophorae or Chalicodomae; nowadays, I have quite got over those terrors. If you do not tease the insect, the thought of hurting you will never occur to it. At the worst, a single specimen, prompted by curiosity rather than anger, will come and hover in front of your face, examining you with some persistency, but employing a buzz as her only threat. Let her be: her scrutiny is quite friendly.

After a few demonstrations, my household were reassured: all, old and young, moved in and out of the arch as though there were nothing unusual about it. My Bees, far from remaining an object of dread, became an object of diversion; every one took pleasure in watching the progress of their ingenious work. I was careful not to divulge the secret to strangers. If any one, coming on business, passed outside the arch while I was standing before the hanging nests, some such brief dialogue as the following would take place:

'So they know you; that's why they don't sting you? '

'They certainly know me. '

'And me? '

'Oh, you; that's another matter! '

Whereupon the intruder would keep at a respectful distance, which was what I wanted.

It is time that we thought of experimenting. The Mason-bees intended for the journey must be marked with a sign whereby I may know them. A solution of gum arabic, thickened with a colouring-powder, red, blue or some other shade, is the material which I use to mark my travellers. The variety in hue will save me from confusing the subjects of my different experiments.

When making my former investigations, I used to mark the Bees at the place where I set them free. For this operation, the insects had to be held in the fingers one after the other; and I was thus exposed to frequent stings, which smarted all the more for being constantly repeated. The consequence was that I was not always quite able to control my fingers and thumbs, to the great detriment of my travellers; for I could easily warp their wing-joints and thus weaken their flight. It was worth while improving the method of operation, both in my own interest and in that of the insect. I must mark the Bee, carry her to a distance and release her, without taking her in my fingers, without once touching her. The experiment was bound to gain by these nice precautions. I will describe the method which I adopted.

The Bee is so much engrossed in her work when she buries her abdomen in the cell and rids herself of her load of pollen, or when she is building, that it is easy, at such times, without alarming her, to mark the upper side of the thorax with a straw dipped in the coloured glue. The insect is not disturbed by that slight touch. It flies off; it returns laden with mortar or pollen. You allow these trips to be repeated until the mark on the thorax is quite dry, which soon happens in the hot sun necessary to the Bee's labours. The next thing is to catch her and imprison her in a paper bag, still without touching her. Nothing could be easier. You place a small test-tube over the Bee engrossed in her work; the insect, on leaving, rushes into it and is thence transferred to the paper bag, which is forthwith closed and placed in the tin box that will serve as a conveyance for the whole party. When releasing the Bees, all you have to do is open the bags. The whole performance is thus effected without once giving that distressing squeeze of the fingers.

Another question remains to be solved before we go further. What time- limit shall I allow for this census of the Bees that return to the nest? Let me explain what I mean. The dot which I have made in the middle of the thorax with a touch of my sticky straw is not very permanent: it merely adheres to the hairs. At the same time, it would have been no more lasting if I had held the insect in my fingers. Now the Bee often brushes her back: she dusts it each time she leaves the galleries; besides, she is always rubbing her coat against the walls of the cell, which she has to enter and to leave each time that she brings honey. A Mason-bee, so smartly dressed at the start, at the end of her work is in rags; her fur is all worn bare and as tattered as a mechanic's overall.

Furthermore, in bad weather, the Mason-bee of the Walls spends the days and nights in one of the cells of her dome, suspended head downwards. The Mason-bee of the Sheds, as long as there are vacant galleries, does very nearly the same: she takes shelter in the galleries, but with her head at the entrance. Once those old habitations are in use, however, and the building of new cells begun, she selects another retreat. In the harmas (The piece of enclosed waste ground on which the author studies his insects in their natural state. Cf. "The Life of the Fly": chapter 1. —Translator's Note.), as I have said elsewhere, are stone heaps, intended for building the surrounding wall. This is where my Chalicodomae pass the night. Piled up promiscuously, both sexes together, they sleep in numerous companies, in crevices between two stones laid closely one on top of the other. Some of these companies number as many as a couple of hundred. The most common dormitory is a narrow groove. Here they all huddle, as far forward as possible, with their backs in the groove. I see some lying flat on their backs, like people asleep. Should bad weather come on, should the sky cloud over, should the north-wind whistle, they do not stir out.

With all these things to take into consideration, I cannot expect my dot on the Bee's thorax to last any length of time. By day, the constant brushing and the rubbing against the partitions of the galleries soon wipe it off; at night, things are worse still, in the narrow sleeping-room where the Mason-bees take refuge by the hundred. After a night spent in the crevice between two stones, it is not advisable to trust to the mark made yesterday. Therefore, the counting of the number of Bees that return to the nest must be taken in hand at once; tomorrow would be too late. And so, as it would be impossible for me to recognize those of my subjects whose dots had disappeared during the night, I will take into account only the Bees that return on the same day.

The question of the rotary machine remains. Darwin advised me to use a circular box with an axle and a handle. I have nothing of the kind in the house. It will be simpler and quite as effective to employ the method of the countryman who tries to lose his Cat by swinging him in a bag. My insects, each one placed by itself in a paper cornet (A cornet is simply the old 'sugar-bag, ' the funnel-shaped paper bag so common on the continent and still used occasionally by small grocers and tobacconists in England. —Translator's Note.) or screw, shall be placed in a tin box; the screws of paper shall be wedged in so as to avoid collisions during the rotation; lastly, the box shall be tied to a cord and I will whirl the whole thing round like a sling. With this contrivance, it will be quite easy to obtain any rate of speed that I wish, any variety of inverse movements that I consider likely to make my captives lose their bearings. I can whirl my sling first in one direction and then in another, turn and turn about; I can slacken or increase the pace; if I like, I can make it describe figures of eight, combined with circles; if I spin on my heels at the same time, I am able to make the process still more complicated by compelling my sling to trace every known curve. That is what I shall do.

On the 2nd of May 1880, I make a white mark on the thorax of ten Mason-bees busied with various tasks: some are exploring the slabs of clay in order to select a site; others are brick-laying; others are garnering stores. When the mark is dry, I catch them and pack them as I have described. I first carry them a quarter of a mile in the opposite direction to the one which I intend to take. A path skirting my house favours this preliminary manoeuvre; I have every hope of being alone when the time comes to make play with my sling. There is a way-side cross at the end; I stop at the foot of the cross. Here I swing my Bees in every direction. Now, while I am making the box describe inverse circles and loops, while I am pirouetting on my heels to achieve the various curves, up comes a woman from the village and stares at me. Oh, how she stares at me, what a look she gives me! At the foot of the cross! Acting in such a silly way! People talked about it. It was sheer witchcraft. Had I not dug up a dead body, only a few days before? Yes, I had been to a prehistoric burial-place, I had taken from it a pair of venerable, well-developed tibias, a set of funerary vessels and a few shoulders of horse, placed there as a viaticum for the great journey. I had done this thing; and people knew it. And now, to crown all, the man of evil reputation is found at the foot of a cross indulging in unhallowed antics.

No matter—and it shows no small courage on my part—the gyrations are duly accomplished in the presence of this unexpected witness. Then I retrace my steps and walk westward of Serignan. I take the least- frequented paths, I cut across country so as, if possible, to avoid a second meeting. It would be the last straw if I were seen opening my paper bags and letting loose my insects! When half-way, to make my experiment more decisive still, I repeat the rotation, in as complicated a fashion as before. I repeat it for the third time at the spot chosen for the release.

I am at the end of a flint-strewn plain, with here and there a scanty curtain of almond-trees and holm-oaks. Walking at a good pace, I have taken thirty minutes to cover the ground in a straight line. The distance therefore is, roughly, two miles. It is a fine day, under a clear sky, with a very light breeze blowing from the north. I sit down on the ground, facing the south, so that the insects may be free to take either the direction of their nest or the opposite one. I let them loose at a quarter past two. When the bags are opened, the Bees, for the most part, circle several times around me and then dart off impetuously in the direction of Serignan, as far as I can judge. It is not easy to watch them, because they fly off suddenly, after going two or three times round my body, a suspicious-looking object which they wish, apparently, to reconnoitre before starting. A quarter of an hour later, my eldest daughter, Antonia, who is on the look-out beside the nests, sees the first traveller arrive. On my return, in the course of the evening, two others come back. Total: three home on the same day, out of ten scattered abroad.

I resume the experiment next morning. I mark ten Mason-bees with red, which will enable me to distinguish them from those who returned on the day before and from those who may still return with the white spot uneffaced. The same precautions, the same rotations, the same localities as on the first occasion; only, I make no rotation on the way, confining myself to swinging my box round on leaving and on arriving. The insects are released at a quarter past eleven. I preferred the forenoon, as this was the busiest time at the works. One Bee was seen by Antonia to be back at the nest by twenty minutes past eleven. Supposing her to be the first let loose, it took her just five minutes to cover the distance. But there is nothing to tell me that it is not another, in which case she needed less. It is the fastest speed that I have succeeded in noting. I myself am back at twelve and, within a short time, catch three others. I see no more during the rest of the evening. Total: four home, out of ten.

The 4th of May is a very bright, calm, warm day, weather highly propitious for my experiments. I take fifty Chalicodomae marked with blue. The distance to be travelled remains the same. I make the first rotation after carrying my insects a few hundred steps in the direction opposite to that which I finally take; in addition, three rotations on the road; a fifth rotation at the place where they are set free. If they do not lose their bearings this time, it will not be for lack of twisting and turning. I begin to open my screws of paper at twenty minutes past nine. It is rather early, for which reason my Bees, on recovering their liberty, remain for a moment undecided and lazy; but, after a short sunbath on a stone where I place them, they take wing. I am sitting on the ground, facing the south, with Serignan on my left and Piolenc on my right. When the flight is not too swift to allow me to perceive the direction taken, I see my released captives disappear to my left. A few, but only a few, go south; two or three go west, or to right of me. I do not speak of the north, against which I act as a screen. All told, the great majority take the left, that is to say, the direction of the nest. The last is released at twenty minutes to ten. One of the fifty travellers has lost her mark in the paper bag. I deduct her from the total, leaving forty-nine.

According to Antonia, who watches the home-coming, the earliest arrivals appeared at twenty-five minutes to ten, say fifteen minutes after the first was set free. By twelve o'clock mid-day, there are eleven back; and, by four o'clock in the evening, seventeen. That ends the census. Total: seventeen, out of forty-nine.

I resolved upon a fourth experiment, on the 14th of May. The weather is glorious, with a light northerly breeze. I take twenty Mason-bees, marked in pink, at eight o'clock in the morning. Rotations at the start, after a preliminary backing in a direction opposite to that which I intend to take; two rotations on the road; a fourth on arriving. All those whose flight I am able to follow with my eyes turn to my left, that is to say, towards Serignan. Yet I had taken care to leave the choice free between the two opposite directions: in particular, I had sent away my Dog, who was on my right. To-day, the Bees do not circle round me: some fly away at once; the others, the greater number, feeling giddy perhaps after the pitching of the journey and the rolling of the sling, alight on the ground a few yards away, seem to wait until they are somewhat recovered and then fly off to the left. I perceived this to be the general flight, whenever I was able to observe at all. I was back at a quarter to ten. Two Bees with pink marks were there before me, of whom one was engaged in building, with her pellet of mortar in her mandibles. By one o'clock in the afternoon there were seven arrivals; I saw no more during the rest of the day. Total: seven out of twenty.

Let us be satisfied with this: the experiment has been repeated often enough, but it does not conclude as Darwin hoped, as I myself hoped, especially after what I had been told about the Cat. In vain, adopting the advice given, do I carry my insects first in the opposite direction to the place at which I intend to release them; in vain, when about to retrace my steps, do I twirl my sling with every complication in the way of whirls and twists that I am able to imagine; in vain, thinking to increase the difficulties, do I repeat the rotation as often as five times over: at the start, on the road, on arriving; it makes no difference: the Mason-bees return; and the proportion of returns on the same day fluctuates between thirty and forty per cent. It goes to my heart to abandon an idea suggested by so famous a man of science and cherished all the more readily inasmuch as I thought it likely to provide a final solution. The facts are there, more eloquent than any number of ingenious views; and the problem remains as mysterious as ever.

In the following year, 1881, I began experimenting again, but in a different way. Hitherto, I had worked on the level. To return to the nest, my lost Bees had only to cross slight obstacles, the hedges and spinneys of the tilled fields. To-day, I propose to add to the difficulties of distance those of the ground to be traversed. Discontinuing all my backing- and whirling-tactics, things which I recognize as useless, I think of releasing my Chalicodomae in the thick of the Serignan Woods. How will they escape from that labyrinth, where, in the early days, I needed a compass to find my way? Moreover, I shall have an assistant with me, a pair of eyes younger than mine and better-fitted to follow my insects' first flight. That immediate start in the direction of the nest has already been repeated very often and is beginning to interest me more than the return itself. A pharmaceutical student, spending a few days with my parents, shall be my eyewitness. With him, I shall feel at ease; science and he are no strangers.

The trip to the woods takes place on the 16th of May. The weather is hot and hints at a coming storm. There is a perceptible breeze from the south, but not enough to upset my travellers. Forty Mason-bees are caught. To shorten the preparations, because of the distance, I do not mark them while they are on the nests; I shall mark them at the starting-point, as I release them. It is the old method, prolific of stings; but I prefer it to-day, in order to save time. It takes me an hour to reach the place. The distance, therefore, allowing for windings, is about three miles.

The site selected must permit me to recognize the direction of the insects' first flight. I choose a clearing in the middle of the copses. All around is a great expanse of dense woods, shutting out the horizon on every side; on the south, in the direction of the nests, a curtain of hills rises to a height of some three hundred feet above the spot at which I stand. The wind is not strong, but it is blowing in the opposite direction to that which my insects will have to take in order to reach their home. I turn my back on Serignan, so that, when leaving my fingers, the Bees, to return to the nest, will be obliged to fly sideways, to right and left of me; I mark the insects and release them one by one. I begin operations at twenty minutes past ten.

One half of the Bees seem rather indolent, flutter about for a while, drop to the ground, appear to recover their spirits and then start off. The other half show greater decision. Although the insects have to fight against the soft wind that is blowing from the south, they make straight for the nest. All go south, after describing a few circles, a few loops, around us. There is no exception in the case of any of those whose departure we are able to follow. The fact is noted by myself and my colleague beyond dispute or doubt. My Mason-bees head for the south as though some compass told them which way the wind was blowing.

I am back at twelve o'clock. None of the strays is at the nest; but, a few minutes later, I catch two. At two o'clock, the number has increased to nine. But now the sky clouds over, the wind freshens and the storm is approaching. We can no longer rely on any further arrivals. Total: nine out of forty, or twenty-two per cent.

The proportion is smaller than in the former cases, when it varied between thirty and forty per cent. Must we attribute this result to the difficulties to be overcome? Can the Mason-bees have lost their way in the maze of the forest? It is safer not to give an opinion: other causes intervened which may have decreased the number of those who returned. I marked the insects at the starting-place; I handled them; and I am not prepared to say that they were all in the best of condition on leaving my stung and smarting fingers. Besides, the sky has become overcast, a storm is imminent. In the month of May, so variable, so fickle, in my part of the world, we can hardly ever count on a whole day of fine weather. A splendid morning is swiftly followed by a fitful afternoon; and my experiments with Mason-bees have often suffered by these variations. All things considered, I am inclined to think that the homeward journey across the forest and the mountain is effected just as readily as across the corn-fields and the plain.

I have one last resource left whereby to try and put my Bees out of their latitude. I will first take them to a great distance; then, describing a wide curve, I will return by another road and release my captives when I am near enough to the village, say, about two miles. A conveyance is necessary, this time. My collaborator of the day in the woods offers me the use of his gig. The two of us set off, with fifteen Mason-bees, along the road to Orange, until we come to the viaduct. Here, on the right, is the straight ribbon of the old Roman road, the Via Domitia. We take it, driving north towards the Uchaux Mountains, the classic home of superb Turonian fossils. We next turn back towards Serignan, by the Piolenc Road. A halt is made by the stretch of country known as Font-Claire, the distance from which to the village is about one mile and five furlongs. The reader can easily follow my route on the ordnance-survey map; and he will see that the loop described measures not far short of five miles and a half.

At the same time, Favier came and joined me at Font-Claire, by the direct road, the one that runs through Piolenc. He brought with him fifteen Mason-bees, intended for purposes of comparison with mine. I am therefore in possession of two sets of insects. Fifteen, marked in pink, have taken the five-mile bend; fifteen, marked in blue, have come by the straight road, the shortest road for returning to the nest. The weather is warm, exceedingly bright and very calm; I could not hope for a better day for my experiment. The insects are given their freedom at mid-day.

At five o'clock, the arrivals number seven of the pink Mason-bees, whom I thought that I had bewildered by a long and circuitous drive, and six of the blue Mason-bees, who came to Font-Claire by the direct route. The two proportions, forty-six and forty per cent., are almost equal; and the slight excess in favour of the insects that went the roundabout way is evidently an accidental result which we need not take into consideration. The bend described cannot have helped them to find their way home; but it has also certainly not hampered them.

There is no need of further proof. The intricate movements of a rotation such as I have described; the obstacle of hills and woods; the pitfalls of a road which moves on, moves back and returns after making a wide circuit: none of these is able to disconcert the Chalicodomae or prevent them from going back to the nest.

I had written to Charles Darwin telling him of my first, negative results, those obtained by swinging the Bees in a box. He expected a success and was much surprised at the failure. Had he had time to experiment with his Pigeons, they would have behaved just like my Bees; the preliminary twirling would not have affected them. The problem called for another method; and what he proposed was this:

'To place the insect within an induction coil, so as to disturb any magnetic or diamagnetic sensibility which it seems just possible that they may possess. '

To treat an insect as you would a magnetic needle and to subject it to the current from an induction coil in order to disturb its magnetism or diamagnetism appeared to me, I must confess, a curious notion, worthy of an imagination in the last ditch. I have but little confidence in our physics, when they pretend to explain life; nevertheless, my respect for the great man would have made me resort to the induction-coils, if I had possessed the necessary apparatus. But my village boasts no scientific resources: if I want an electric spark, I am reduced to rubbing a sheet of paper on my knees. My physics cupboard contains a magnet; and that is about all. When this penury was realised, another method was suggested, simpler than the first and more certain in its results, as Darwin himself considered:

'To make a very thin needle into a magnet; then breaking it into very short pieces, which would still be magnetic, and fastening one of these pieces with some cement on the thorax of the insects to be experimented on. I believe that such a little magnet, from its close proximity to the nervous system of the insect, would affect it more than would the terrestrial currents. '

There is still the same idea of turning the insect into a sort of bar magnet. The terrestrial currents guide it when returning to the nest. It becomes a living compass which, withdrawn from the action of the earth by the proximity of a loadstone, loses its sense of direction. With a tiny magnet fastened on its thorax, parallel with the nervous system and more powerful than the terrestrial magnetism by reason of its comparative nearness, the insect will lose its bearings. Naturally, in setting down these lines, I take shelter behind the mighty reputation of the learned begetter of the idea. It would not be accepted as serious coming from a humble person like myself. Obscurity cannot afford these audacious theories.

The experiment seems easy; it is not beyond the means at my disposal. Let us attempt it. I magnetise a very fine needle by rubbing it with my bar magnet; I retain only the slenderest part, the point, some five or six millimetres long. (. 2 to. 23 inch. —Translator's Note.) This broken piece is a perfect magnet: it attracts and repels another magnetised needle hanging from a thread. I am a little puzzled as to the best way to fasten it on the insect's thorax. My assistant of the moment, the pharmaceutical student, requisitions all the adhesives in his laboratory. The best is a sort of cerecloth which he prepares specially with a very fine material. It possesses the advantage that it can be softened at the bowl of one's pipe when the time comes to operate out of doors.

I cut out of this cerecloth a small square the size of the Bee's thorax; and I insert the magnetised point through a few threads of the material. All that we now have to do is to soften the gum a little and then dab the thing at once on the Mason-bee's back, so that the broken needle runs parallel with the spine. Other engines of the same kind are prepared and due note taken of their poles, so as to enable me to point the south pole at the insect's head in some cases and at the opposite end in others.

My assistant and I begin by rehearsing the performance; we must have a little practice before trying the experiment away from home. Besides, I want to see how the insect will behave in its magnetic harness. I take a Mason-bee at work in her cell, which I mark. I carry her to my study, at the other end of the house. The magnetised outfit is fastened on the thorax; and the insect is let go. The moment she is free, the Bee drops to the ground and rolls about, like a mad thing, on the floor of the room. She resumes her flight, flops down again, turns over on her side, on her back, knocks against the things in her way, buzzes noisily, flings herself about desperately and ends by darting through the open window in headlong flight.

What does it all mean? The magnet appears to have a curious effect on my patient's system! What a fuss she makes! How terrified she is! The Bee seemed utterly distraught at losing her bearings under the influence of my knavish tricks. Let us go to the nests and see what happens. We have not long to wait: my insect returns, but rid of its magnetic tackle. I recognize it by the traces of gum that still cling to the hair of the thorax. It goes back to its cell and resumes its labours.

Always on my guard when searching the unknown, unwilling to draw conclusions before weighing the arguments for and against, I feel doubt creeping in upon me with regard to what I have seen. Was it really the magnetic influence that disturbed my Bee so strangely? When she struggled and kicked on the floor, fighting wildly with both legs and wings, when she fled in terror, was she under the sway of the magnet fastened on her back? Can my appliance have thwarted the guiding influence of the terrestrial currents on her nervous system? Or was her distress merely the result of an unwonted harness? This is what remains to be seen and that without delay.

I construct a new apparatus, but provide it with a short straw in place of the magnet. The insect carrying it on its back rolls on the ground, kicks and flings herself about like the first, until the irksome contrivance is removed, taking with it a part of the fur on the thorax. The straw produces the same effects as the magnet, in other words, magnetism had nothing to do with what happened. My invention, in both cases alike, is a cumbrous tackle of which the Bee tries to rid herself at once by every possible means. To look to her for normal actions so long as she carries an apparatus, magnetized or not, upon her back is the same as expecting to study the natural habits of a Dog after tying a kettle to his tail.

The experiment with the magnet is impracticable. What would it tell us if the insect consented to it? In my opinion, it would tell us nothing. In the matter of the homing instinct, a magnet would have no more influence than a bit of straw.

CHAPTER 5
THE STORY OF MY CATS.

If this swinging-process fails entirely when its object is to make the insect lose its bearings, what influence can it have upon the Cat? Is the method of whirling the animal round in a bag, to prevent its return, worthy of confidence? I believed in it at first, so close- allied was it to the hopeful idea suggested by the great Darwin. But my faith is now shaken: my experience with the insect makes me doubtful of the Cat. If the former returns after being whirled, why should not the latter? I therefore embark upon fresh experiments.

And, first of all, to what extent does the Cat deserve his reputation of being able to return to the beloved home, to the scenes of his amorous exploits on the tiles and in the hay-lofts? The most curious facts are told of his instinct; children's books on natural history abound with feats that do the greatest credit to his prowess as a pilgrim. I do not attach much importance to these stories: they come from casual observers, uncritical folk given to exaggeration. It is not everybody who can talk about animals correctly. When some one not of the craft gets on the subject and says to me, 'Such or such an animal is black, ' I begin by finding out if it does not happen to be white; and many a time the truth is discovered in the converse proposition. Men come to me and sing the praises of the Cat as a travelling-expert. Well and good: we will now look upon the Cat as a poor traveller. And that would be the extent of my knowledge if I had only the evidence of books and of people unaccustomed to the scruples of scientific examination. Fortunately, I am acquainted with a few incidents that will stand the test of my incredulity. The Cat really deserves his reputation as a discerning pilgrim. Let us relate these incidents.

One day—it was at Avignon—there appeared upon the garden-wall a wretched-looking Cat, with matted coat and protruding ribs, so thin that his back was a mere jagged ridge. He was mewing with hunger. My children, at that time very young, took pity on his misery. Bread soaked in milk was offered him at the end of a reed. He took it. And the mouthfuls succeeded one another to such good purpose that he was sated and went off, heedless of the 'Puss! Puss! ' of his compassionate friends. Hunger returned; and the starveling reappeared in his wall- top refectory. He received the same fare of bread soaked in milk, the same soft words. He allowed himself to be tempted. He came down from the wall. The children were able to stroke his back. Goodness, how thin he was!

It was the great topic of conversation. We discussed it at table: we would tame the vagabond, we would keep him, we would make him a bed of hay. It was a most important matter: I can see to this day, I shall always see the council of rattleheads deliberating on the Cat's fate. They were not satisfied until the savage animal remained. Soon he grew into a magnificent Tom. His large round head, his muscular legs, his reddish fur, flecked with darker patches, reminded one of a little jaguar. He was christened Ginger because of his tawny hue. A mate joined him later, picked up in almost similar circumstances. Such was the origin of my series of Gingers, which I have retained for little short of twenty years through the vicissitudes of my various removals.

The first of these removals took place in 1870. A little earlier, a minister who has left a lasting memory in the University, that fine man, Victor Duruy (Jean Victor Duruy (1811-1894), author of a number of historical works, including a well-known "Histoire des Romains", and minister of public instruction under Napoleon III. from 1863 to 1869. Cf. "The Life of the Fly": chapter 20. — Translator's Note.), had instituted classes for the secondary education of girls. This was the beginning, as far as was then possible, of the burning question of to-day. I very gladly lent my humble aid to this labour of light. I was put to teach physical and natural science. I had faith and was not sparing of work, with the result that I rarely faced a more attentive or interested audience. The days on which the lessons fell were red- letter days, especially when the lesson was botany and the table disappeared from view under the treasures of the neighbouring conservatories.

That was going too far. In fact, you can see how heinous my crime was: I taught those young persons what air and water are; whence the lightning comes and the thunder; by what device our thoughts are transmitted across the seas and continents by means of a metal wire; why fire burns and why we breathe; how a seed puts forth shoots and how a flower blossoms: all eminently hateful things in the eyes of some people, whose feeble eyes are dazzled by the light of day.

The little lamp must be put out as quickly as possible and measures taken to get rid of the officious person who strove to keep it alight.

The scheme was darkly plotted with the old maids who owned my house and who saw the abomination of desolation in these new educational methods. I had no written agreement to protect me. The bailiff appeared with a notice on stamped paper. It baldly informed that I must move out within four weeks from date, failing which the law would turn my goods and chattels into the street. I had hurriedly to provide myself with a dwelling. The first house which we found happened to be at Orange. Thus was my exodus from Avignon effected.

We were somewhat anxious about the moving of the Cats. We were all of us attached to them and should have thought it nothing short of criminal to abandon the poor creatures, whom we had so often petted, to distress and probably to thoughtless persecution. The shes and the kittens would travel without any trouble: all you have to do is to put them in a basket; they will keep quiet on the journey. But the old Tom-cats were a serious problem. I had two: the head of the family, the patriarch; and one of his descendants, quite as strong as himself. We decided to take the grandsire, if he consented to come, and to leave the grandson behind, after finding him a home.

My friend Dr. Loriol offered to take charge of the forsaken one. The animal was carried to him at nightfall in a closed hamper. Hardly were we seated at the evening-meal, talking of the good fortune of our Tom- cat, when we saw a dripping mass jump through the window. The shapeless bundle came and rubbed itself against our legs, purring with happiness. It was the Cat.

I learnt his story next day. On arriving at Dr. Loriol's, he was locked up in a bedroom. The moment he saw himself a prisoner in the unfamiliar room, he began to jump about wildly on the furniture, against the window-panes, among the ornaments on the mantelpiece, threatening to make short work of everything. Mme. Loriol was frightened by the little lunatic; she hastened to open the window; and the Cat leapt out among the passers-by. A few minutes later, he was back at home. And it was no easy matter: he had to cross the town almost from end to end; he had to make his way through a long labyrinth of crowded streets, amid a thousand dangers, including first boys and next dogs; lastly—and this perhaps was an even more serious obstacle—he had to pass over the Sorgue, a river running through Avignon. There were bridges at hand, many, in fact; but the animal, taking the shortest cut, had used none of them, bravely jumping into the water, as its streaming fur showed.

I had pity on the poor Cat, so faithful to his home. We agreed to do our utmost to take him with us. We were spared the worry: a few days later, he was found lying stiff and stark under a shrub in the garden. The plucky animal had fallen a victim to some stupid act of spite. Some one had poisoned him for me. Who? It is not likely that it was a friend!

There remained the old Cat. He was not indoors when we started; he was prowling round the hay-lofts of the neighbourhood. The carrier was promised an extra ten francs if he brought the Cat to Orange with one of the loads which he had still to convey. On his last journey he brought him stowed away under the driver's seat. I scarcely knew my old Tom when we opened the moving prison in which he had been confined since the day before. He came out looking a most alarming beast, scratching and spitting, with bristling hair, bloodshot eyes, lips white with foam. I thought him mad and watched him closely for a time. I was wrong: it was merely the fright of a bewildered animal. Had there been trouble with the carrier when he was caught? Did he have a bad time on the journey? History is silent on both points. What I do know is that the very nature of the Cat seemed changed: there was no more friendly purring, no more rubbing against our legs; nothing but a wild expression and the deepest gloom. Kind treatment could not soothe him. For a few weeks longer, he dragged his wretched existence from corner to corner; then, one day, I found him lying dead in the ashes on the hearth. Grief, with the help of old age, had killed him. Would he have gone back to Avignon, had he had the strength? I would not venture to affirm it. But, at least, I think it very remarkable that an animal should let itself die of home-sickness because the infirmities of age prevent it from returning to its old haunts.

What the patriarch could not attempt, we shall see another do, over a much shorter distance, I admit. A fresh move is resolved upon, that I may have, at length, the peace and quiet essential to my work. This time, I hope that it will be the last. I leave Orange for Serignan.

The family of Gingers has been renewed: the old ones have passed away, new ones have come, including a full-grown Tom, worthy in all respects of his ancestors. He alone will give us some difficulty; the others, the babies and the mothers, can be removed without trouble. We put them into baskets. The Tom has one to himself, so that the peace may be kept. The journey is made by carriage, in company with my family. Nothing striking happens before our arrival. Released from their hampers, the females inspect the new home, explore the rooms one by one; with their pink noses they recognize the furniture: they find their own seats, their own tables, their own arm-chairs; but the surroundings are different. They give little surprised miaows and questioning glances. A few caresses and a saucer of milk allay all their apprehensions; and, by the next day, the mother Cats are acclimatised.

It is a different matter with the Tom. We house him in the attics, where he will find ample room for his capers; we keep him company, to relieve the weariness of captivity; we take him a double portion of plates to lick; from time to time, we place him in touch with some of his family, to show him that he is not alone in the house; we pay him a host of attentions, in the hope of making him forget Orange. He appears, in fact, to forget it: he is gentle under the hand that pets him, he comes when called, purrs, arches his back. It is well: a week of seclusion and kindly treatment have banished all notions of returning. Let us give him his liberty. He goes down to the kitchen, stands by the table like the others, goes out into the garden, under the watchful eye of Aglae, who does not lose sight of him; he prowls all around with the most innocent air. He comes back. Victory! The Tom-cat will not run away.

Next morning:

'Puss! Puss! '

Not a sign of him! We hunt, we call. Nothing. Oh, the hypocrite, the hypocrite! How he has tricked us! He has gone, he is at Orange. None of those about me can believe in this venturesome pilgrimage. I declare that the deserter is at this moment at Orange mewing outside the empty house.

Aglae and Claire went to Orange. They found the Cat, as I said they would, and brought him back in a hamper. His paws and belly were covered with red clay; and yet the weather was dry, there was no mud. The Cat, therefore, must have got wet crossing the Aygues torrent; and the moist fur had kept the red earth of the fields through which he passed. The distance from Serignan to Orange, in a straight line, is four and a half miles. There are two bridges over the Aygues, one above and one below that line, some distance away. The Cat took neither the one nor the other: his instinct told him the shortest road and he followed that road, as his belly, covered with red mud, proved. He crossed the torrent in May, at a time when the rivers run high; he overcame his repugnance to water in order to return to his beloved home. The Avignon Tom did the same when crossing the Sorgue.

The deserter was reinstated in his attic at Serignan. He stayed there for a fortnight; and at last we let him out. Twenty-four hours had not elapsed before he was back at Orange. We had to abandon him to his unhappy fate. A neighbour living out in the country, near my former house, told me that he saw him one day hiding behind a hedge with a rabbit in his mouth. Once no longer provided with food, he, accustomed to all the sweets of a Cat's existence, turned poacher, taking toll of the farm-yards round about my old home. I heard no more of him. He came to a bad end, no doubt: he had become a robber and must have met with a robber's fate.

The experiment has been made and here is the conclusion, twice proved. Full-grown Cats can find their way home, in spite of the distance and their complete ignorance of the intervening ground. They have, in their own fashion, the instinct of my Mason-bees. A second point remains to be cleared up, that of the swinging motion in the bag. Are they thrown out of their latitude by this stratagem, are or they not? I was thinking of making some experiments, when more precise information arrived and taught me that it was not necessary. The first who acquainted me with the method of the revolving bag was telling the story told him by a second person, who repeated the story of a third, a story related on the authority of a fourth; and so on. None had tried it, none had seen it for himself. It is a tradition of the country-side. One and all extol it as an infallible method, without, for the most part, having attempted it. And the reason which they give for its success is, in their eyes, conclusive. If, say they, we ourselves are blind-folded and then spin round for a few seconds, we no longer know where we are. Even so with the Cat carried off in the darkness of the swinging bag. They argue from man to the animal, just as others argue from the animal to man: a faulty method in either case, if there really be two distinct psychic worlds.

The belief would not be so deep-rooted in the peasant's mind, if facts had not from time to time confirmed it. But we may assume that, in successful cases, the Cats made to lose their bearings were young and unemancipated animals. With those neophytes, a drop of milk is enough to dispel the grief of exile. They do not return home, whether they have been whirled in a bag or not. People have thought it as well to subject them to the whirling operation by way of an additional precaution; and the method has received the credit of a success that has nothing to do with it. In order to test the method properly, it should have been tried on a full-grown Cat, a genuine Tom.

I did in the end get the evidence which I wanted on this point. Intelligent and trustworthy people, not given to jumping to conclusions, have told me that they have tried the trick of the swinging bag to keep Cats from returning to their homes. None of them succeeded when the animal was full-grown. Though carried to a great distance, into another house, and subjected to a conscientious series of revolutions, the Cat always came back. I have in mind more particularly a destroyer of the Goldfish in a fountain, who, when transported from Serignan to Piolenc, according to the time-honoured method, returned to his fish; who, when carried into the mountain and left in the woods, returned once more. The bag and the swinging round proved of no avail; and the miscreant had to be put to death. I have verified a fair number of similar instances, all under most favourable conditions. The evidence is unanimous: the revolving motion never keeps the adult Cat from returning home. The popular belief, which I found so seductive at first, is a country prejudice, based upon imperfect observation. We must, therefore, abandon Darwin's idea when trying to explain the homing of the Cat as well as of the Mason-bee.

CHAPTER 6
THE RED ANTS.

The Pigeon transported for hundreds of miles is able to find his way back to his Dove-cot; the Swallow, returning from his winter quarters in Africa, crosses the sea and once more takes possession of the old nest. What guides them on these long journeys? Is it sight? An observer of supreme intelligence, one who, though surpassed by others in the knowledge of the stuffed animal under a glass case, is almost unrivalled in his knowledge of the live animal in its wild state, Toussenel (Alphonse Toussenel (1803-1885), the author of a number of interesting and valuable works on ornithology. — Translator's Note.), the admirable writer of "L'Esprit des betes", speaks of sight and meteorology as the Carrier-pigeon's guides:

'The French bird, ' he says, 'knows by experience that the cold weather comes from the north, the hot from the south, the dry from the east and the wet from the west. That is enough meteorological knowledge to tell him the cardinal points and to direct his flight. The Pigeon taken in a closed basket from Brussels to Toulouse has certainly no means of reading the map of the route with his eyes; but no one can prevent him from feeling, by the warmth of the atmosphere, that he is pursuing the road to the south. When restored to liberty at Toulouse, he already knows that the direction which he must follow to regain his Dove-cot is the direction of the north. Therefore he wings straight in that direction and does not stop until he nears those latitudes where the mean temperature is that of the zone which he inhabits. If he does not find his home at the first onset, it is because he has borne a little too much to the right or to the left. In any case, it takes him but a few hours' search in an easterly or westerly direction to correct his mistake. '

The explanation is a tempting one when the journey is taken north and south; but it does not apply to a journey east and west, on the same isothermal line. Besides, it has this defect, that it does not admit of generalization. One cannot talk of sight and still less of the influence of a change of climate when a Cat returns home, from one end of a town to the other, threading his way through a labyrinth of streets and alleys which he sees for the first time. Nor is it sight that guides my Mason-bees, especially when they are let loose in the thick of a wood. Their low flight, eight or nine feet above the ground, does not allow them to take a panoramic view nor to gather the lie of the land. What need have they of topography? Their hesitation is short-lived: after describing a few narrow circles around the experimenter, they start in the direction of the nest, despite the cover of the forest, despite the screen of a tall chain of hills which they cross by mounting the slope at no great height from the ground. Sight enables them to avoid obstacles, without giving them a general idea of their road. Nor has meteorology aught to do with the case: the climate has not varied in those few miles of transit. My Mason-bees have not learnt from any experience of heat, cold, dryness and damp: an existence of a few weeks' duration does not allow of this. And, even if they knew all about the four cardinal points, there is no difference in climate between the spot where their nest lies and the spot at which they are released; so that does not help them to settle the direction in which they are to travel.

To explain these many mysteries, we are driven therefore to appeal to yet another mystery, that is to say, a special sense denied to mankind. Charles Darwin, whose weighty authority no one will gainsay, arrives at the same conclusion. To ask if the animal be not impressed by the terrestrial currents, to enquire if it be not influenced by the close proximity of a magnetic needle: what is this but the recognition of a magnetic sense? Do we possess a similar faculty? I am speaking, of course, of the magnetism of the physicists and not of the magnetism of the Mesmers and Cagliostros. Assuredly we possess nothing remotely like it. What need would the mariner have of a compass, were he himself a compass?

And this is what the great scientist acknowledges: a special sense, so foreign to our organism that we are not able to form a conception of it, guides the Pigeon, the Swallow, the Cat, the Mason-bee and a host of others when away from home. Whether this sense be magnetic or no I will not take upon myself to decide; I am content to have helped, in no small degree, to establish its existence. A new sense added to our number: what an acquisition, what a source of progress! Why are we deprived of it? It would have been a fine weapon and of great service in the struggle for life. If, as is contended, the whole of the animal kingdom, including man, is derived from a single mould, the original cell, and becomes self-evolved in the course of time, favouring the best-endowed and leaving the less well-endowed to perish, how comes it that this wonderful sense is the portion of a humble few and that it has left no trace in man, the culminating achievement of the zoological progression? Our precursors were very ill-advised to let so magnificent an inheritance go: it was better worth keeping than a vertebra of the coccyx or a hair of the moustache.

Does not the fact that this sense has not been handed down to us point to a flaw in the pedigree? I submit the little problem to the evolutionists; and I should much like to know what their protoplasm and their nucleus have to say to it.

Is this unknown sense localized in a particular part of the Wasp and the Bee? Is it exercised by means of a special organ? We immediately think of the antennae. The antennae are what we always fall back upon when the insect's actions are not quite clear to us; we gladly put down to them whatever is most necessary to our arguments. For that matter, I had plenty of fairly good reasons for suspecting them of containing the sense of direction. When the Hairy Ammophila (A Sand- wasp who hunts the Grey Worm, or Caterpillar of the Turnip-moth, to serve as food for her grubs. For other varieties of the Ammophila, cf. "Insect Life": chapter 15. —Translator's Note.) is searching for the Grey Worm, it is with her antennae, those tiny fingers continually fumbling at the soil, that she seems to recognize the presence of the underground prey. Could not those inquisitive filaments, which seem to guide the insect when hunting, also guide it when travelling? This remained to be seen; and I did see.

I took some Mason-bees and amputated their antennae with the scissors, as closely as I could. These maimed ones were then carried to a distance and released. They returned to the nest with as little difficulty as the others. I once experimented in the same way with the largest of our Cerceres (Cerceris tuberculata) (Another Hunting Wasp, who feeds her young on Weevils. Cf. "Insect Life": chapters 4 and 5. — Translator's Note.); and the Weevil-huntress returned to her galleries. This rids us of one hypothesis: the sense of direction is not exercised by the antennae. Then where is its seat? I do not know.

What I do know is that the Mason-bees without antennae, though they go back to the cells, do not resume work. They persist in flying in front of their masonry, they alight on the clay cup, they perch on the rim of the cell and there, seemingly pensive and forlorn, stand for a long time contemplating the work which will never be finished; they go off, they come back, they drive away any importunate neighbour, but they fetch and carry no more honey or mortar. The next day, they do not appear. Deprived of her tools, the worker loses all heart in her task. When the Mason-bee is building, the antennae are constantly feeling, fumbling and exploring, superintending, as it were, the finishing touches given to the work. They are her instruments of precision; they represent the builder's compasses, square, level and plumb-line.

Hitherto my experiments have been confined to the females, who are much more faithful to the nest by virtue of their maternal responsibilities. What would the males do if they were taken from home? I have no great confidence in these swains who, for a few days, form a tumultuous throng outside the nests, wait for the females to emerge, quarrel for their possession, amid endless brawls, and then disappear when the works are in full swing. What care they, I ask myself, about returning to the natal nest rather than settling elsewhere, provided that they find some recipient for their amatory declarations? I was mistaken: the males do return to the nest. It is true that, in view of their lack of strength, I did not subject them to a long journey: about half a mile or so. Nevertheless, this represented to them a distant expedition, an unknown country; for I do not see them go on long excursions. By day, they visit the nests or the flowers in the garden; at night, they take refuge in the old galleries or in the interstices of the stone-heaps in the harmas.

The same nests are frequented by two Osmia-bees (Osmia tricornis and Osmia Latreillii), who build their cells in the galleries left at their disposal by the Chalicodomae. The most numerous is the first, the Three-horned Osmia. It was a splendid opportunity to try and discover to what extent the sense of direction may be regarded as general in the Bees and Wasps; and I took advantage of it. Well, the Osmiae (Osmia tricornis), both male and female, can find their way back to the nest. My experiments were made very quickly, with small numbers and over short distances; but the results agreed so closely with the others that I was convinced. All told, the return to the nest, including my earlier attempts, was verified in the case of four species: the Chalicodoma of the Sheds, the Chalicodoma of the Walls, the Three-horned Osmia and the Great or Warted Cerceris (Cerceris tuberculata). ("Insect Life": chapter 19. —Translator's Note.) Shall I generalize without reserve and allow all the Hymenoptera (The Hymenoptera are an order of insects having four membranous wings and include the Bees, Wasps, Ants, Saw-flies and Ichneumon-flies. — Translator's Note.) this faculty of finding their way in unknown country? I shall do nothing of the kind; for here, to my knowledge, is a contradictory and very significant result.

Among the treasures of my harmas-laboratory, I place in the first rank an Ant-hill of Polyergus rufescens, the celebrated Red Ant, the slave- hunting Amazon. Unable to rear her family, incapable of seeking her food, of taking it even when it is within her reach, she needs servants who feed her and undertake the duties of housekeeping. The Red Ants make a practice of stealing children to wait on the community. They ransack the neighbouring Ant-hills, the home of a different species; they carry away nymphs, which soon attain maturity in the strange house and become willing and industrious servants.

When the hot weather of June and July sets in, I often see the Amazons leave their barracks of an afternoon and start on an expedition. The column measures five or six yards in length. If nothing worthy of attention be met upon the road, the ranks are fairly well maintained; but, at the first suspicion of an Ant-hill, the vanguard halts and deploys in a swarming throng, which is increased by the others as they come up hurriedly. Scouts are sent out; the Amazons recognize that they are on a wrong track; and the column forms again. It resumes its march, crosses the garden-paths, disappears from sight in the grass, reappears farther on, threads its way through the heaps of dead leaves, comes out again and continues its search. At last, a nest of Black Ants is discovered. The Red Ants hasten down to the dormitories where the nymphs lie and soon emerge with their booty. Then we have, at the gates of the underground city, a bewildering scrimmage between the defending blacks and the attacking reds. The struggle is too unequal to remain indecisive. Victory falls to the reds, who race back to their abode, each with her prize, a swaddled nymph, dangling from her mandibles. The reader who is not acquainted with these slave-raiding habits would be greatly interested in the story of the Amazons. I relinquish it, with much regret: it would take us too far from our subject, namely, the return to the nest.

The distance covered by the nymph-stealing column varies: it all depends on whether Black Ants are plentiful in the neighbourhood. At times, ten or twenty yards suffice; at others, it requires fifty, a hundred or more. I once saw the expedition go beyond the garden. The Amazons scaled the surrounding wall, which was thirteen feet high at that point, climbed over it and went on a little farther, into a cornfield. As for the route taken, this is a matter of indifference to the marching column. Bare ground, thick grass, a heap of dead leaves or stones, brickwork, a clump of shrubs: all are crossed without any marked preference for one sort of road rather than another.

What is rigidly fixed is the path home, which follows the outward track in all its windings and all its crossings, however difficult. Laden with their plunder, the Red Ants return to the nest by the same road, often an exceedingly complicated one, which the exigencies of the chase compelled them to take originally. They repass each spot which they passed at first; and this is to them a matter of such imperative necessity that no additional fatigue nor even the gravest danger can make them alter the track.

Let us suppose that they have crossed a thick heap of dead leaves, representing to them a path beset with yawning gulfs, where every moment some one falls, where many are exhausted as they struggle out of the hollows and reach the heights by means of swaying bridges, emerging at last from the labyrinth of lanes. No matter: on their return, they will not fail, though weighed down with their burden, once more to struggle through that weary maze. To avoid all this fatigue, they would have but to swerve slightly from the original path, for the good, smooth road is there, hardly a step away. This little deviation never occurs to them.

I came upon them one day when they were on one of their raids. They were marching along the inner edge of the stone-work of the garden- pond, where I have replaced the old batrachians by a colony of Gold- fish. The wind was blowing very hard from the north and, taking the column in flank, sent whole rows of the Ants flying into the water. The fish hurried up; they watched the performance and gobbled up the drowning insects. It was a difficult bit; and the column was decimated before it had passed. I expected to see the return journey made by another road, which would wind round and avoid the fatal cliff. Not at all. The nymph-laden band resumed the parlous path and the Goldfish received a double windfall: the Ants and their prizes. Rather than alter its track, the column was decimated a second time.

It is not easy to find the way home again after a distant expedition, during which there have been various sorties, nearly always by different paths; and this difficulty makes it absolutely necessary for the Amazons to return by the same road by which they went. The insect has no choice of route, if it would not be lost on the way: it must come back by the track which it knows and which it has lately travelled. The Processionary Caterpillars, when they leave their nest and go to another branch, on another tree, in search of a type of leaf more to their taste, carpet the course with silk and are able to return home by following the threads stretched along their road. This is the most elementary method open to the insect liable to stray on its excursions: a silken path brings it home again. The Processionaries, with their unsophisticated traffic-laws, are very different from the Mason-bees and others, who have a special sense to guide them.

The Amazon, though belonging to the Hymenopteron clan, herself possesses rather limited homing-faculties, as witness her compulsory return by her former trail. Can she imitate, to a certain extent, the Processionaries' method, that is to say, does she leave, along the road traversed, not a series of conducting threads, for she is not equipped for that work, but some odorous emanation, for instance some formic scent, which would allow her to guide herself by means of the olfactory sense? This view is pretty generally accepted. The Ants, people say, are guided by the sense of smell; and this sense of smell appears to have its seat in the antennae, which we see in continual palpitation. It is doubtless very reprehensible, but I must admit that the theory does not inspire me with overwhelming enthusiasm. In the first place, I have my suspicions about a sense of smell seated in the antennae: I have given my reasons before; and, next, I hope to prove by experiment that the Red Ants are not guided by a scent of any kind.

To lie in wait for my Amazons, for whole afternoons on end, often unsuccessfully, meant taking up too much of my time. I engaged an assistant whose hours were not so much occupied as mine. It was my grand-daughter Lucie, a little rogue who liked to hear my stories of the Ants. She had been present at the great battle between the reds and blacks and was much impressed by the rape of the long-clothes babies. Well-coached in her exalted functions, very proud of already serving that august lady, Science, my little Lucie would wander about the garden, when the weather seemed propitious, and keep an eye on the Red Ants, having been commissioned to reconnoitre carefully the road to the pillaged Ant-hill. She had given proof of her zeal; I could rely upon it.

One day, while I was spinning out my daily quota of prose, there came a banging at my study-door:

'It's I, Lucie! Come quick: the reds have gone into the blacks' house. Come quick! '

'And do you know the road they took? '

'Yes, I marked it. '

'What! Marked it? How? '

'I did what Hop-o'-my-Thumb did: I scattered little white stones along the road. '

I hurried out. Things had happened as my six-year-old colleague said. Lucie had secured her provision of pebbles in advance and, on seeing the Amazon regiment leave barracks, had followed them step by step and placed her stones at intervals along the road covered. The Ants had made their raid and were beginning to return along the track of tell- tale pebbles. The distance to the nest was about a hundred paces, which gave me time to make preparations for an experiment previously contemplated.

I take a big broom and sweep the track for about a yard across. The dusty particles on the surface are thus removed and replaced by others. If they were tainted with any odorous effluvia, their absence will throw the Ants off the track. I divide the road, in this way, at four different points, a few feet a part.

The column arrives at the first section. The hesitation of the Ants is evident. Some recede and then return, only to recede once more; others wander along the edge of the cutting; others disperse sideways and seem to be trying to skirt the unknown country. The head of the column, at first closed up to a width of a foot or so, now scatters to three or four yards. But fresh arrivals gather in their numbers before the obstacle; they form a mighty array, an undecided horde. At last, a few Ants venture into the swept zone and others follow, while a few have meantime gone ahead and recovered the track by a circuitous route. At the other cuttings, there are the same halts, the same hesitations; nevertheless, they are crossed, either in a straight line or by going round. In spite of my snares, the Ants manage to return to the nest; and that by way of the little stones.

The result of the experiment seems to argue in favour of the sense of smell. Four times over, there are manifest hesitations wherever the road is swept. Though the return takes place, nevertheless, along the original track, this may be due to the uneven work of the broom, which has left certain particles of the scented dust in position. The Ants who went round the cleared portion may have been guided by the sweepings removed to either side. Before, therefore, pronouncing judgment for or against the sense of smell, it were well to renew the experiment under better conditions and to remove everything containing a vestige of scent.

A few days later, when I have definitely decided on my plan, Lucie resumes her watch and soon comes to tell me of a sortie. I was counting on it, for the Amazons rarely miss an expedition during the hot and sultry afternoons of June and July, especially when the weather threatens storm. Hop-o'-my-Thumb's pebbles once more mark out the road, on which I choose the point best-suited to my schemes.

A garden-hose is fixed to one of the feeders of the pond; the sluice is opened; and the Ants' path is cut by a continuous torrent, two or three feet wide and of unlimited length. The sheet of water flows swiftly and plentifully at first, so as to wash the ground well and remove anything that may possess a scent. This thorough washing lasts for nearly a quarter of an hour. Then, when the Ants draw near, returning from the plunder, I let the water flow more slowly and reduce its depth, so as not to overtax the strength of the insects. Now we have an obstacle which the Amazons must surmount, if it is absolutely necessary for them to follow the first trail.

This time, the hesitation lasts long and the stragglers have time to come up with the head of the column. Nevertheless, an attempt is made to cross the torrent by means of a few bits of gravel projecting above the water; then, failing to find bottom, the more reckless of the Ants are swept off their feet and, without loosing hold of their prizes, drift away, land on some shoal, regain the bank and renew their search for a ford. A few straws borne on the waters stop and become so many shaky bridges on which the Ants climb. Dry olive-leaves are converted into rafts, each with its load of passengers. The more venturesome, partly by their own efforts, partly by good luck, reach the opposite bank without adventitious aid. I see some who, dragged by the current to one or the other bank, two or three yards off, seem very much concerned as to what they shall do next. Amid this disorder, amid the dangers of drowning, not one lets go her booty. She would not dream of doing so: death sooner than that! In a word, the torrent is crossed somehow or other along the regular track.

The scent of the road cannot be the cause of this, it seems to me, for the torrent not only washed the ground some time beforehand but also pours fresh water on it all the time that the crossing is taking place. Let us now see what will happen when the formic scent, if there really be one on the trail, is replaced by another, much stronger odour, one perceptible to our own sense of smell, which the first is not, at least not under present conditions.

I wait for a third sortie and, at one point in the road taken by the Ants, rub the ground with some handfuls of freshly gathered mint. I cover the track, a little farther on, with the leaves of the same plant. The Ants, on their return, cross the section over which the mint was rubbed without apparently giving it a thought; they hesitate in front of the section heaped up with leaves and then go straight on.

After these two experiments, first with the torrent of water which washes away all traces of smell from the ground and then with the mint which changes the smell, I think that we are no longer at liberty to quote scent as the guide of the Ants that return to the nest by the road which they took at starting. Further tests will tell us more about it.

Without interfering with the soil, I now lay across the track some large sheets of paper, newspapers, keeping them in position with a few small stones. In front of this carpet, which completely alters the appearance of the road, without removing any sort of scent that it may possess, the Ants hesitate even longer than before any of my other snares, including the torrent. They are compelled to make manifold attempts, reconnaissances to right and left, forward movements and repeated retreats, before venturing altogether into the unknown zone. The paper straits are crossed at last and the march resumed as usual.

Another ambush awaits the Amazons some distance farther on. I have divided the track by a thin layer of yellow sand, the ground itself being grey. This change of colour alone is enough for a moment to disconcert the Ants, who again hesitate in the same way, though not for so long, as they did before the paper. Eventually, this obstacle is overcome like the others.

As neither the stretch of sand nor the stretch of paper got rid of any scented effluvia with which the trail may have been impregnated, it is patent that, as the Ants hesitated and stopped in the same way as before, they find their way not by sense of smell, but really and truly by sense of sight; for, every time that I alter the appearance of the track in any way whatever—whether by my destructive broom, my streaming water, my green mint, my paper carpet or my golden sand—the returning column calls a halt, hesitates and attempts to account for the changes that have taken place. Yes, it is sight, but a very dull sight, whose horizon is altered by the shifting of a few bits of gravel. To this short sight, a strip of paper, a bed of mint-leaves, a layer of yellow sand, a stream of water, a furrow made by the broom, or even lesser modifications are enough to transform the landscape; and the regiment, eager to reach home as fast as it can with its loot, halts uneasily on beholding this unfamiliar scenery. If the doubtful zones are at length passed, it is due to the fact that fresh attempts are constantly being made to cross the doctored strips and that at last a few Ants recognize well-known spots beyond them. The others, relying on their clearer-sighted sisters, follow.

Sight would not be enough, if the Amazon had not also at her service a correct memory for places. The memory of an Ant! What can that be? In what does it resemble ours? I have no answers to these questions; but a few words will enable me to prove that the insect has a very exact and persistent recollection of places which it has once visited. Here is something which I have often witnessed. It sometimes happens that the plundered Ant-hill offers the Amazons a richer spoil than the invading column is able to carry away. Or, again, the region visited is rich in Ant-hills. Another raid is necessary, to exploit the site thoroughly. In such cases, a second expedition takes place, sometimes on the next day, sometimes two or three days later. This time, the column does no reconnoitring on the way: it goes straight to the spot known to abound in nymphs and travels by the identical path which it followed before. It has sometimes happened that I have marked with small stones, for a distance of twenty yards, the road pursued a couple of days earlier and have then found the Amazons proceeding by the same route, stone by stone:

'They will go first here and then there, ' I said, according to the position of the guide-stones.

And they would, in fact, go first here and then there, skirting my line of pebbles, without any noticeable deviation.

Can one believe that odoriferous emanations diffused along the route are going to last for several days? No one would dare to suggest it. It must, therefore, be sight that directs the Amazons, sight assisted by a memory for places. And this memory is tenacious enough to retain the impression until the next day and later; it is scrupulously faithful, for it guides the column by the same path as on the day before, across the thousand irregularities of the ground.

How will the Amazon behave when the locality is unknown to her? Apart from topographical memory, which cannot serve her here, the region in which I imagine her being still unexplored, does the Ant possess the Mason-bee's sense of direction, at least within modest limits, and is she able thus to regain her Ant-hill or her marching column?

The different parts of the garden are not all visited by the marauding legions to the same extent: the north side is exploited by preference, doubtless because the forays in that direction are more productive. The Amazons, therefore, generally direct their troops north of their barracks; I seldom see them in the south. This part of the garden is, if not wholly unknown, at least much less familiar to them than the other. Having said that, let us observe the conduct of the strayed Ant.

I take up my position near the Ant-hill; and, when the column returns from the slave-raid, I force an Ant to step on a leaf which I hold out to her. Without touching her, I carry her two or three paces away from her regiment: no more than that, but in a southerly direction. It is enough to put her astray, to make her lose her bearings entirely. I see the Amazon, now replaced on the ground, wander about at random, still, I need hardly say, with her booty in her mandibles; I see her hurry away from her comrades, thinking that she is rejoining them; I see her retrace her steps, turn aside again, try to the right, try to the left and grope in a host of directions, without succeeding in finding her whereabouts. The pugnacious, strong-jawed slave-hunter is utterly lost two steps away from her party. I have in mind certain strays who, after half an hour's searching, had not succeeded in recovering the route and were going farther and farther from it, still carrying the nymph in their teeth.

What became of them? What did they do with their spoil? I had not the patience to follow those dull- witted marauders to the end.

Let us repeat the experiment, but place the Amazon to the north. After more or less prolonged hesitations, after a search now in this direction, now in that, the Ant succeeds in finding her column. She knows the locality.

Here, of a surety, is a Hymenopteron deprived of that sense of direction which other Hymenoptera enjoy. She has in her favour a memory for places and nothing more. A deviation amounting to two or three of our strides is enough to make her lose her way and to keep her from returning to her people, whereas miles across unknown country will not foil the Mason-bee. I expressed my surprise, just now, that man was deprived of a wonderful sense wherewith certain animals are endowed. The enormous distance between the two things compared might furnish matter for discussion. In the present case, the distance no longer exists: we have to do with two insects very near akin, two Hymenoptera. Why, if they issue from the same mould, has one a sense which the other has not, an additional sense, constituting a much more overpowering factor than the structural details? I will wait until the evolutionists condescend to give me a valid reason.

To return to this memory for places whose tenacity and fidelity I have just recognized: to what degree does it consent to retain impressions? Does the Amazon require repeated journeys in order to learn her geography, or is a single expedition enough for her? Are the line followed and the places visited engraved on her memory from the first? The Red Ant does not lend herself to the tests that might furnish the reply: the experimenter is unable to decide whether the path followed by the expeditionary column is being covered for the first time, nor is it in his power to compel the legion to adopt this or that different road. When the Amazons go out to plunder the Ant-hills, they take the direction which they please; and we are not allowed to interfere with their march. Let us turn to other Hymenoptera for information.

I select the Pompili, whose habits we shall study in detail in a later chapter. (For the Wasp known as the Pompilus, or Ringed Calicurgus, cf. "The Life and Love of the Insect", by J. Henri Fabre, translated by Alexander Teixeira de Mattos: chapter 12. — Translator's Note.) They are hunters of Spiders and diggers of burrows. The game, the food of the coming larva, is first caught and paralysed; the home is excavated afterwards. As the heavy prey would be a grave encumbrance to the Wasp in search of a convenient site, the Spider is placed high up, on a tuft of grass or brushwood, out of the reach of marauders, especially Ants, who might damage the precious morsel in the lawful owner's absence. After fixing her booty on the verdant pinnacle, the Pompilus casts around for a favourable spot and digs her burrow. During the process of excavation, she returns from time to time to her Spider; she nibbles at the prize, feels, touches it here and there, as though taking stock of its plumpness and congratulating herself on the plentiful provender; then she returns to her burrow and goes on digging. Should anything alarm or distress her, she does not merely inspect her Spider: she also brings her a little closer to her work-yard, but never fails to lay her on the top of a tuft of verdure. These are the manoeuvres of which I can avail myself to gauge the elasticity of the Wasp's memory.

While the Pompilus is at work on the burrow, I seize the prey and place it in an exposed spot, half a yard away from its original position. The Pompilus soon leaves the hole to enquire after her booty and goes straight to the spot where she left it. This sureness of direction, this faithful memory for places can be explained by repeated previous visits. I know nothing of what has happened beforehand. Let us take no notice of this first expedition; the others will be more conclusive. For the moment, the Pompilus, without the least hesitation, finds the tuft of grass whereon her prey was lying. Then come marches and counter-marches upon that tuft, minute explorations and frequent returns to the exact spot where the Spider was deposited. At last, convinced that the prize is no longer there, the Wasp makes a leisurely survey of the neighbourhood, feeling the ground with her antennae as she goes. The Spider is descried in the exposed spot where I had placed her. Surprise on the part of the Pompilus, who goes forward and then suddenly steps back with a start:

'Is it alive? ' she seems to ask. 'Is it dead? Is it really my Spider? Let us be wary! '

The hesitation does not last long: the huntress grabs her victim, drags her backwards and places her, still high up, on a second tuft of herbage, two or three steps away from the first. She then goes back to the burrow and digs for a while. For the second time, I remove the

Spider and lay her at some distance, on the bare ground. This is the moment to judge of the Wasp's memory. Two tufts of grass have served as temporary resting-places for the game. The first, to which she returned with such precision, the Wasp may have learnt to know by a more or less thorough examination, by reiterated visits that escaped my eye; but the second has certainly made but a slight impression on her memory. She adopted it without any studied choice; she stopped there just long enough to hoist her Spider to the top; she saw it for the first time and saw it hurriedly, in passing. Is that rapid glance enough to provide an exact recollection? Besides, there are now two localities to be modelled in the insect's memory: the first shelf may easily be confused with the second. To which will the Pompilus go?

We shall soon find out: here she comes, leaving the burrow to pay a fresh visit to the Spider. She runs straight to the second tuft, where she hunts about for a long time for her absent prey. She knows that it was there, when last seen, and not elsewhere; she persists in looking for it there and does not once think of going back to the first perch. The first tuft of grass no longer counts; the second alone interests her. And then the search in the neighbourhood begins again.

On finding her game on the bare spot where I myself have placed it, the Pompilus quickly deposits the Spider on a third tuft of grass; and the experiment is renewed. This time, the Pompilus hurries to the third tuft when she comes to look after her Spider; she hurries to it without hesitation, without confusing it in any way with the first two, which she scorns to visit, so sure is her memory. I do the same thing a couple of times more; and the insect always returns to the last perch, without worrying about the others. I stand amazed at the memory of that pigmy. She need but catch a single hurried glimpse of a spot that differs in no wise from a host of others in order to remember it quite well, notwithstanding the fact that, as a miner relentlessly pursuing her underground labours, she has other matters to occupy her mind. Could our own memory always vie with hers? It is very doubtful. Allow the Red Ant the same sort of memory; and her peregrinations, her returns to the nest by the same road are no longer difficult to explain.

Tests of this kind have furnished me with some other results worthy of mention. When convinced, by untiring explorations, that her prey is no longer on the tuft where she laid it, the Pompilus, as we were saying, looks for it in the neighbourhood and finds it pretty easily, for I am careful to put it in an exposed place. Let us increase the difficulty to some extent. I dig the tip of my finger into the ground and lay the Spider in the little hole thus obtained, covering her with a tiny leaf. Now the Wasp, while in quest of her lost prey, happens to walk over this leaf, to pass it again and again without suspecting that the Spider lies beneath, for she goes and continues her vain search farther off. Her guide, therefore is not scent, but sight. Nevertheless, she is constantly feeling the ground with her antennae. What can be the function of those organs? I do not know, although I assert that they are not olfactory organs. The Ammophila, in search of her Grey Worm, had already led me to make the same assertion; I now obtain an experimental proof which seems to me decisive. I would add that the Pompilus has very short sight: often she passes within a couple of inches of her Spider without seeing her.

CHAPTER 7
SOME REFLECTIONS UPON INSECT PSYCHOLOGY.

The laudator temperis acti is out of favour just now: the world is on the move. Yes, but sometimes it moves backwards. When I was a boy, our twopenny textbooks told us that man was a reasoning animal; nowadays, there are learned volumes to prove to us that human reason is but a higher rung in the ladder whose foot reaches down to the bottommost depths of animal life. There is the greater and the lesser; there are all the intermediary rounds; but nowhere does it break off and start afresh. It begins with zero in the glair of a cell and ascends until we come to the mighty brain of a Newton. The noble faculty of which we were so proud is a zoological attribute. All have a larger or smaller share of it, from the live atom to the anthropoid ape, that hideous caricature of man.

It always struck me that those who held this levelling theory made facts say more than they really meant; it struck me that, in order to obtain their plain, they were lowering the mountain-peak, man, and elevating the valley, the animal. Now this levelling of theirs needed proofs, to my mind; and, as I found none in their books, or at any rate only doubtful and highly debatable ones, I did my own observing, in order to arrive at a definite conviction; I sought; I experimented.

To speak with any certainty, it behoves us not to go beyond what we really know. I am beginning to have a passable acquaintance with insects, after spending some forty years in their company. Let us question the insect, then: not the first that comes along, but the most gifted, the Hymenopteron. I am giving my opponents every advantage. Where will they find a creature more richly endowed with talent? It would seem as though, in creating it, nature had delighted in bestowing the greatest amount of industry upon the smallest body of matter. Can the bird, wonderful architect that it is, compare its work with that masterpiece of higher geometry, the edifice of the Bee? The Hymenopteron rivals man himself. We build towns, the Bee erects cities; we have servants, the Ant has hers; we rear domestic animals, she rears her sugar-yielding insects; we herd cattle, she herds her milch-cows, the Aphides; we have abolished slavery, whereas she continues her nigger-traffic.

Well, does this superior, this privileged being reason? Reader, do not smile: this is a most serious matter, well worthy of our consideration. To devote our attention to animals is to plunge at once into the vexed question of who we are and whence we come. What, then, passes in that little Hymenopteron brain? Has it faculties akin to ours, has it the power of thought? What a problem, if we could only solve it; what a chapter of psychology, if we could only write it! But, at our very first questionings, the mysterious will rise up, impenetrable: we may be convinced of that. We are incapable of knowing ourselves; what will it be if we try to fathom the intellect of others? Let us be content if we succeed in gleaning a few grains of truth.

What is reason? Philosophy would give us learned definitions. Let us be modest and keep to the simplest: we are only treating of animals. Reason is the faculty that connects the effect with its cause and directs the act by conforming it to the needs of the accidental. Within these limits, are animals capable of reasoning? Are they able to connect a 'because' with a 'why' and afterwards to regulate their behaviour accordingly? Are they able to change their line of conduct when faced with an emergency?

History has but few data likely to be of use to us here; and those which we find scattered in various authors are seldom able to withstand a severe examination. One of the most remarkable of which I know is supplied by Erasmus Darwin, in his book entitled "Zoonomia. " It tells of a Wasp that has just caught and killed a big Fly. The wind is blowing; and the huntress, hampered in her flight by the great area presented by her prize, alights on the ground to amputate the abdomen, the head and the wings; she flies away, carrying with her only the thorax, which gives less hold to the wind. If we keep to the bald facts, this does, I admit, give a semblance of reason. The Wasp appears to grasp the relation between cause and effect. The effect is the resistance experienced in the flight; the cause is the dimensions of the prey contending with the air. Hence the logical conclusion: those dimensions must be lessened; the abdomen, the head and, above all, the wings must be chopped off; and the resistance will be decreased. (I would gladly, if I were able, cancel some rather hasty lines which I allowed myself to pen in the first volume of these "Souvenirs" but scripta manent. All that I can do is to make amends now, in this note, for the error into which I fell. Relying on Lacordaire, who quotes this instance from Erasmus Darwin in his own "Introduction a l'entomologie", I believed that a Sphex was given as the heroine of the story. How could I do otherwise, not having the original text in front of me? How could I suspect that an entomologist of Lacordaire's standing should be capable of such a blunder as to substitute a Sphex for a Common Wasp? Great was my perplexity, in the face of this evidence! A Sphex capturing a Fly was an impossibility; and I blamed the British scientist accordingly. But what insect was it that Erasmus Darwin saw? Calling logic to my aid, I declared that it was a Wasp; and I could not have hit the mark more truly. Charles Darwin, in fact, informed me afterwards that his grandfather wrote 'a Wasp' in his "Zoonomia. " Though the correction did credit to my intelligence, I none the less deeply regretted my mistake, for I had uttered suspicions of the observer's powers of discernment, unjust suspicions which the translator's inaccuracy led me into entertaining. May this note serve to mitigate the harshness of the strictures provoked by my overtaxed credulity! I do not scruple to attack ideas which I consider false; but Heaven forfend that I should ever attack those who uphold them! — Author's Note.)

But does this concatenation of ideas, rudimentary though it be, really take place within the insect's brain? I am convinced of the contrary; and my proofs are unanswerable. In the first volume of these "Souvenirs" (Cf. "Insect Life": chapter 9. —Translator's Note.), I demonstrated by experiment that Erasmus Darwin's Wasp was but obeying her instinct, which is to cut up the captured game and to keep only the most nourishing part, the thorax. Whether the day be perfectly calm or whether the wind blow, whether she be in the shelter of a dense thicket or in the open, I see the Wasp proceed to separate the succulent from the tough; I see her reject the legs, the wings, the head and the abdomen, retaining only the breast as pap for her larvae. Then what value has this dissection as an argument in favour of the insect's reasoning-powers when the wind blows? It has no value at all, for it would take place just the same in absolutely calm weather. Erasmus Darwin jumped too quickly to his conclusion, which was the outcome of his mental bias and not of the logic of things. If he had first enquired into the Wasp's habits, he would not have brought forward as a serious argument an incident which had no connection with the important question of animal reason.

I have reverted to this case to show the difficulties that beset the man who confines himself to casual observations, however carefully carried out. One should never rely upon a lucky chance, which may not occur again. We must multiply our observations, check them one with the other; we must create incidents, looking into preceding ones, finding out succeeding ones and working out the relation between them all: then and not till then, with extreme caution, are we entitled to express a few views worthy of credence. Nowhere do I find data collected under such conditions; for which reason, however much I might wish it, it is impossible for me to bring the evidence of others in support of the few conclusions which I myself have formed.

My Mason-bees, with their nests hanging on the walls of the arch which I have mentioned, lent themselves to continuous experiment better than any other Hymenopteron. I had them there, at my house, under my eyes, at all hours of the day, as long as I wished. I was free to follow their actions in full detail and to carry out successfully any experiment, however long. Moreover, their numbers allowed me to repeat my attempts until I was perfectly convinced. The Mason-bees, therefore, shall supply me with the materials for this chapter also.

A few words, before I begin, about the works. The Mason-bee of the Sheds utilizes, first of all, the old galleries of the clay nest, a part of which she good-naturedly abandons to two Osmiae, her free tenants: the Three-horned Osmia and Latreille's Osmia. These old corridors, which save labour, are in great demand; but there are not many vacant, as the more precocious Osmiae have already taken possession of most of them; and therefore the building of new cells soon begins. These cells are cemented to the surface of the nest, which thus increases in thickness every year. The edifice of cells is not built all at once: mortar and honey alternate repeatedly. The masonry starts with a sort of little swallow's nest, a half-cup or thimble, whose circumference is completed by the wall against which it rests. Picture the cup of an acorn cut in two and stuck to the surface of the nest: there you have the receptacle in a stage sufficiently advanced to take a first instalment of honey.

The Bee thereupon leaves the mortar and busies herself with harvesting. After a few foraging-trips, the work of building is resumed; and some new rows of bricks raise the edge of the basin, which becomes capable of receiving a larger stock of provisions. Then comes another change of business: the mason once more becomes a harvester. A little later, the harvester is again a mason; and these alternations continue until the cell is of the regulation height and holds the amount of honey required for the larva's food. Thus come, turn and turn about, more or less numerous according to the occupation in hand, journeys to the dry and barren path, where the cement is gathered and mixed, and journeys to the flowers, where the Bee's crop is crammed with honey and her belly powdered with pollen.

At last comes the time for laying. We see the Bee arrive with a pellet of mortar. She gives a glance at the cell to enquire if everything is in order; she inserts her abdomen; and the egg is laid. Then and there the mother seals up the home: with her pellet of cement she closes the orifice and manages so well with the material that the lid receives its permanent form at this first sitting; it has only to be thickened and strengthened with fresh layers, a work which is less urgent and will be done by and by. What does appear to be an urgent necessity is the closing of the cell immediately after the egg has been religiously deposited therein, so that there may be no danger from evilly-disposed visitors during the mother's absence. The Bee must have serious reasons for thus hurrying on the closing of the cell. What would happen if, after laying her egg, she left the house open and went to the cement-pit to fetch the wherewithal to block the door? Some thief might drop in and substitute her own egg for the Mason-bee's. We shall see that our suspicions are not uncalled-for. One thing is certain, that the Mason never lays without having in her mandibles the pellet of mortar required for the immediate construction of the lid of the nest. The precious egg must not for a single instant remain exposed to the cupidity of marauders.

To these particulars I will add a few general observations which will make what follows easier to understand. So long as its circumstances are normal, the insect's actions are calculated most rationally in view of the object to be attained. What could be more logical, for instance, than the devices employed by the Hunting Wasp when paralysing her prey (Cf. "Insect Life": chapters 3 to 12 and 15 to 17. — Translator's Note.) so that it may keep fresh for her larva, while in no wise imperilling that larva's safety? It is preeminently rational; we ourselves could think of nothing better; and yet the Wasp's action is not prompted by reason. If she thought out her surgery, she would be our superior. It will never occur to anybody that the creature is able, in the smallest degree, to account for its skilful vivisections. Therefore, so long as it does not depart from the path mapped out for it, the insect can perform the most sagacious actions without entitling us in the least to attribute these to the dictates of reason.

What would happen in an emergency? Here we must distinguish carefully between two classes of emergency, or we shall be liable to grievous error. First, in accidents occurring in the course of the insect's occupation at the moment. In these circumstances, the creature is capable of remedying the accident; it continues, under a similar form, its actual task; it remains, in short; in the same psychic condition. In the second case, the accident is connected with a more remote occupation; it relates to a completed task with which, under normal conditions, the insect is no longer concerned. To meet this emergency, the creature would have to retrace its psychic course; it would have to do all over again what it has just finished, before turning its attention to anything else. Is the insect capable of this? Will it be able to leave the present and return to the past? Will it decide to hark back to a task that is much more pressing than the one on which it was engaged? If it did all this, then we should really have evidence of a modicum of reason. The question shall be settled by experiment.

We will begin by taking a few incidents that come under the first heading. A Mason-bee has finished the initial layer of the covering of the cell. She has gone in search of a second pellet of mortar wherewith to strengthen her work. In her absence, I prick the lid with a needle and widen the hole thus made, until it is half the size of the opening. The insect returns and repairs the damage. It was originally engaged on the lid and is merely continuing its work in mending that lid.

A second is still at her first row of bricks. The cell as yet is no more than a shallow cup, containing no provisions. I make a big hole in the bottom of the cup and the Bee hastens to stop the breach. She was busy building and turned aside a moment to do more building. Her repairs are the continuation of the work on which she was engaged.

A third has laid her egg and closed the cell. While she is gone in search of a fresh supply of cement to strengthen the door, I make a large aperture immediately below the lid, too high up to allow the honey to escape. The insect, on arriving with its mortar intended for a different task, sees its broken jar and soon puts the damage right. I have rarely witnessed such a sensible performance. Nevertheless, all things considered, let us not be too lavish of our praises. The insect was busy closing up. On its return, it sees a crack, representing in its eyes a bad join which it had overlooked; it completes its actual task by improving the join.

The conclusion to be drawn from these three instances, which I select from a large number of others, more or less similar, is that the insect is able to cope with emergencies, provided that the new action be not outside the course of its actual work at the moment. Shall we say then that reason directs it? Why should we? The insect persists in the same psychic course, it continues its action, it does what it was doing before, it corrects what to it appears but a careless flaw in the work of the moment.

Here, moreover, is something which would change our estimate entirely, if it ever occurred to us to look upon these repaired breaches as a work dictated by reason. Let us turn to the second class of emergency referred to above: let us imagine, first, cells similar to those in the second experiment, that is to say, only half-finished, in the form of a shallow cup, but already containing honey. I make a hole in the bottom, through which the provisions ooze and run to waste. Their owners are harvesting. Let us imagine, on the other hand, cells very nearly finished and almost completely provisioned. I perforate the bottom in the same way and let out the honey, which drips through gradually. The owners of these are building.

Judging by what has gone before, the reader will perhaps expect to see immediate repairs, urgent repairs, for the safety of the future larva is at stake. Let him dismiss any such illusion: more and more journeys are undertaken, now in quest of food, now in quest of mortar; but not one of the Mason-bees troubles about the disastrous breach. The harvester goes on harvesting; the busy bricklayer proceeds with her next row of bricks, as though nothing out of the way had happened. Lastly, if the injured cells are high enough and contain enough provisions, the Bee lays her eggs, puts a door to the house and passes on to another house, without doing aught to remedy the leakage of the honey. Two or three days later, those cells have lost all their contents, which now form a long trail on the surface of the nest.

Is it through lack of intelligence that the Bee allows her honey to go to waste? May it not rather be through helplessness? It might happen that the sort of mortar which the Mason has at her disposal will not set on the edges of a hole that is sticky with honey. The honey may prevent the cement from adjusting itself to the orifice, in which case the insect's inertness would merely be resignation to an irreparable evil. Let us look into the matter before drawing inferences. With my forceps, I deprive the Bee of her pellet of mortar and apply it to the hole whence the honey is escaping. My attempt at repairing meets with the fullest success, though I do not pretend to compete with the Mason in dexterity. For a piece of work done by a man's hand it is quite creditable. My dab of mortar fits nicely into the mutilated wall; it hardens as usual; and the escape of honey ceases. This is quite satisfactory. What would it be had the work been done by the insect, equipped with its tools of exquisite precision? When the Mason-bee refrains, therefore, this is not due to helplessness on her part, nor to any defect in the material employed.

Another objection presents itself. We are going too far perhaps in admitting this concatenation of ideas in the insect's mind, in expecting it to argue that the honey is running away because the cell has a hole in it and that to save it from being wasted the hole must be stopped. So much logic perhaps exceeds the powers of its poor little brain. Then, again, the hole is not seen; it is hidden by the honey trickling through. The cause of that stream of honey is an unknown cause; and to trace the loss of the liquid home to that cause, to the hole in the receptacle, is too lofty a piece of reasoning for the insect.

A cell in the rudimentary cup-stage and containing no provisions has a hole, three or four millimetres (. 11 to. 15 inch. —Translator's Note.) wide, made in it at the bottom. A few moments later, this orifice is stopped by the Mason. We have already witnessed a similar patching. The insect, having finished, starts foraging. I reopen the hole at the same place. The pollen runs through the aperture and falls to the ground as the Bee is rubbing off her first load in the cell. The damage is undoubtedly observed. When plunging her head into the cup to take stock of what she has stored, the Bee puts her antennae into the artificial hole: she sounds it, she explores it, she cannot fail to perceive it.

I see the two feelers quivering outside the hole. The insect notices the breach in the wall: that is certain. It flies off. Will it bring back mortar from its present journey to repair the injured jar as it did just now?

Not at all. It returns with provisions, it disgorges its honey, it rubs off its pollen, it mixes the material. The sticky and almost solid mass fills up the opening and oozes through with difficulty. I roll a spill of paper and free the hole, which remains open and shows daylight distinctly in both directions. I sweep the place clear over and over again, whenever this becomes necessary because new provisions are brought; I clean the opening sometimes in the Bee's absence, sometimes in her presence, while she is busy mixing her paste. The unusual happenings in the warehouse plundered from below cannot escape her any more than the ever-open breach at the bottom of the cell. Nevertheless, for three consecutive hours, I witness this strange sight: the Bee, full of active zeal for the task in hand, omits to plug this vessel of the Danaides. She persists in trying to fill her cracked receptacle, whence the provisions disappear as soon as stored away. She constantly alternates between builder's and harvester's work; she raises the edges of the cell with fresh rows of bricks; she brings provisions which I continue to abstract, so as to leave the breach always visible. She makes thirty-two journeys before my eyes, now for mortar, now for honey, and not once does she bethink herself of stopping the leakage at the bottom of her jar.

At five o'clock in the evening, the works cease. They are resumed on the morrow. This time, I neglect to clean out my artificial orifice and leave the victuals gradually to ooze out by themselves. At length, the egg is laid and the door sealed up, without anything being done by the Bee in the matter of the disastrous breach. And yet to plug the hole were an easy matter for her: a pellet of her mortar would suffice. Besides, while the cup was still empty, did she not instantly close the hole which I had made? Why are not those early repairs of hers repeated? It clearly shows the creature's inability to retrace the course of its actions, however slightly. At the time of the first breach, the cup was empty and the insect was laying the first rows of bricks. The accident produced through my agency concerned the part of the work which occupied the Bee at the actual moment; it was a flaw in the building, such as can occur naturally in new courses of masonry, which have not had time to harden. In correcting that flaw, the Mason did not go outside her usual work.

But, once the provisioning begins, the cup is finished for good and all; and, come what may, the insect will not touch it again. The harvester will go on harvesting, though the pollen trickle to the ground through the drain. To plug the hole would imply a change of occupation of which the insect is incapable for the moment. It is the honey's turn and not the mortar's. The rule upon this point is invariable. A moment comes, presently, when the harvesting is interrupted and the masoning resumed. The edifice must be raised a storey higher. Will the Bee, once more a builder, mixing fresh cement, now attend to the leakage at the bottom? No more than before. What occupies her at present is the new floor, whose brickwork would be repaired at once, if it sustained a damage; but the bottom storey is too old a part of the business, it is ancient history; and the worker will not put a further touch to it, even though it be in serious danger.

For the rest, the present and the following storeys will all have the same fate. Carefully watched by the insect as long as they are in process of building, they are forgotten and allowed to go to ruin once they are actually built. Here is a striking instance: in a cell which has attained its full height, I make a window, almost as large as the natural opening, and place it about half-way up, above the honey. The Bee brings provisions for some time longer and then lays her egg. Through my big window, I see the egg deposited on the victuals. The insect next works at the cover, to which it gives the finishing touches with a series of little taps, administered with infinite care, while the breach remains yawning. On the lid, it scrupulously stops up every pore that could admit so much as an atom; but it leaves the great opening that places the house at the mercy of the first-comer. It goes to that breach repeatedly, puts in its head, examines it, explores it with its antennae, nibbles the edges of it. And that is all. The mutilated cell shall stay as it is, with never a dab of mortar. The threatened part dates too far back for the Bee to think of troubling about it.

I have said enough, I think, to show the insect's mental incapacity in the presence of the accidental. This incapacity is confirmed by renewing the test, an essential condition of all good experiments; therefore my notes are full of examples similar to the one which I have just described. To relate them would be mere repetition; I pass them over for the sake of brevity.

The renewal of a test is not sufficient: we must also vary our test. Let us, then, examine the insect's intelligence from another point of view, that of the introduction of foreign bodies into the cell. The Mason-bee is a housekeeper of scrupulous cleanliness, as indeed are all the Hymenoptera. Not a spot of dirt is suffered in her honey-pot; not a grain of dust is permitted on the surface of her mixture. And yet, while the jar is open, the precious Bee-bread is exposed to accidents. The workers in the cells above may inadvertently drop a little mortar into the lower cells; the owner herself, when working at enlarging the jar, runs the risk of letting a speck of cement fall into the provisions. A Gnat, attracted by the smell, may come and be caught in the honey; brawls between neighbours who are getting into each other's way may send some dust flying thither. All this refuse has to disappear and that quickly, lest afterwards the larva should find coarse fare under its delicate mandibles. Therefore the Mason- bees must be able to cleanse the cell of any foreign body. And, in point of fact, they are well able to do so.

I place on the surface of the honey five or six bits of straw a millimetre in length. (. 039 inch. —Translator's Note.) Great astonishment on the part of the returning insect. Never before have so many sweepings accumulated in its warehouse. The Bee picks out the bits of straw, one by one, to the very last, and each time goes and gets rid of them at a distance. The effort is out of all proportion to the work: I see the Bee soar above the nearest plane-tree, to a height of thirty feet, and fly away beyond it to rid herself of her burden, a mere atom. She fears lest she should litter the place by dropping her bit of straw on the ground, under the nest. A thing like that must be carried very far away.

I place upon the honey-paste a Mason-bee's egg which I myself saw laid in an adjacent cell. The Bee picks it out and throws it away at a distance, as she did with the straws just now. There are two inferences to be drawn from this, both extremely interesting. In the first place, that precious egg, for whose future the Bee labours so indefatigably, becomes a valueless, cumbersome, hateful thing when it belongs to another. Her own egg is everything; the egg of her next door neighbour is nothing. It is flung on the dust-heap like any bit of rubbish. The individual, so zealous on behalf of her family, displays an abominable indifference for the rest of her kind. Each one for himself. In the second place, I ask myself, without as yet being able to find an answer to my question, how certain parasites go to work to give their larva the benefit of the provisions accumulated by the Mason-bee. If they decide to lay their egg on the victuals in the open cell, the Bee, when she sees it, will not fail to cast it out; if they decide to lay after the owner, they cannot do so, for she blocks up the door as soon as her laying is done. This curious problem must be reserved for future investigation. (Cf. "The Life of the Fly": chapters 2 to 4; also later chapters in the present volume. — Translator's Note.)

Lastly, I stick into the paste a bit of straw nearly an inch long and standing well out above the rim of the cell. The insect extracts it by dint of great efforts, dragging it away from one side; or else, with the help of its wings, it drags it from above. It darts away with the honey-smeared straw and gets rid of it at a distance, after flying over the plane-tree.

This is where things begin to get complicated. I have said that, when the time comes for laying, the Mason-bee arrives with a pellet of mortar wherewith immediately to make a door to the house. The insect, with its front legs resting on the rim, inserts its abdomen in the cell; it has the mortar ready in its mouth. Having laid the egg, it comes out and turns round to block the door. I wave it away for a second, at the same time planting my straw as before, a straw sticking out nearly a centimetre. (. 39 inch. — Translator's Note.) What will the Bee do? Will she, who is scrupulous in ridding the home of the least mote of dust, extract this beam, which would certainly prove the larva's undoing by interfering with its growth? She could, for just now we saw her drag out and throw away, at a distance, a similar beam.

She could and she doesn't. She closes the cell, cements the lid, seals up the straw in the thickness of the mortar. More journeys are taken, not a few, in search of the cement required to strengthen the cover. Each time, the mason applies the material with the most minute care, while giving the straw not a thought. In this way, I obtain, one after the other, eight closed cells whose lids are surmounted by my mast, a bit of protruding straw. What evidence of obtuse intelligence!

This result is deserving of attentive consideration. At the moment when I am inserting my beam, the insect has its mandibles engaged: they are holding the pellet of mortar intended for the blocking-operation. As the extracting-tool is not free, the extraction does not take place. I expected to see the Bee relinquish her mortar and then proceed to remove the encumbrance. A dab of mortar more or less is not a serious business. I had already noticed that it takes my Mason-bees a journey of three or four minutes to collect one. The pollen-expeditions last longer, a matter of ten or fifteen minutes. To drop her pellet, grab the straw with her mandibles, now disengaged, remove it and gather a fresh supply of cement would entail a loss of five minutes at most. The Bee decides differently. She will not, she cannot relinquish her pellet; and she uses it. No matter that the larva will perish by this untimely trowelling: the moment has come to wall up the door; the door is walled up. Once the mandibles are free, the extraction could be attempted, at the risk of wrecking the lid. But the Bee does nothing of the sort: she keeps on fetching mortar; and the lid is religiously finished.

We might go on to say that, if the Bee were obliged to depart in quest of fresh mortar after dropping the first to withdraw the straw, she would leave the egg unguarded and that this would be an extreme measure which the mother cannot bring herself to adopt. Then why does she not place the pellet on the rim of the cell? The mandibles, now free, would remove the beam; the pellet would be taken up again at once; and everything would go to perfection. But no: the insect has its mortar and, come what may, employs it on the work for which it was intended.

If any one sees a rudiment of reason in this Hymenopteron intelligence, he has eyes that are more penetrating than mine. I see nothing in it all but an invincible persistence in the act once begun. The cogs have gripped; and the rest of the wheels must follow. The mandibles are fastened on the pellet of mortar; and the idea, the wish to unfasten them will never occur to the insect until the pellet has fulfilled its purpose. And here is a still greater absurdity: the plugging once begun is very carefully finished with fresh relays of mortar! Exquisite attention is paid to a closing-up which is henceforth useless; no attention at all to the dangerous beam. O little gleams of reason that are said to enlighten the animal, you are very near the darkness, you are naught!

Another and still more eloquent fact will finally convince whoso may yet be doubting. The ration of honey stored up in a cell is evidently measured by the needs of the coming larva. There is neither too much nor too little. How does the Bee know when the proper quantity is reached? The cells are more or less constant in dimension, but they are not filled completely, only to about two-thirds of their height. A large space is therefore left empty; and the victualler has to judge of the moment when the surface of the mess has attained the right level. The honey being perfectly opaque, its depth is not apparent. I have to use a sounding-rod when I want to gauge the contents of the jar; and I find, on the average, that the honey reaches a depth of ten millimetres. (. 39 inch. —Translator's Note.) The Bee has not this resource; she has sight, which may enable her to estimate the full section from the empty section. This presupposes the possession of a somewhat geometric eye, capable of measuring the third of a distance. If the insect did it by Euclid, that would be very brilliant of it. What a magnificent proof in favour of its little intellect: a Chalicodoma with a geometrician's eye, able to divide a straight line into three equal parts! This is worth looking into seriously.

I take five cells, which are only partly provisioned, and empty them of their honey with a wad of cotton held in my forceps. From time to time, as the Bee brings new provisions, I repeat the cleansing-process, sometimes clearing out the cell entirely, sometimes leaving a thin layer at the bottom. I do not observe any pronounced hesitation on the part of my plundered victims, even though they surprise me at the moment when I am draining the jar; they continue their work with quiet industry. Sometimes, two or three threads of cotton remain clinging to the walls of the cells: the Bees remove them carefully and dart away to a distance, as usual, to get rid of them. At last, a little sooner or a little later, the egg is laid and the lid fastened on.

I break open the five closed cells. In one, the egg has been laid on three millimetres of honey (. 117 inch. —Translator's Note.); in two, on one millimetre (. 039 inch. —Translator's Note.); and, in the two others, it is placed on the side of the receptacle drained of all its contents, or, to be more accurate, having only the glaze, the varnish left by the friction of the honey-covered cotton.

The inference is obvious: the Bee does not judge of the quantity of honey by the elevation of the surface; she does not reason like a geometrician, she does not reason at all. She accumulates so long as she feels within her the secret impulse that prompts her to go on collecting until the victualling is completed; she ceases to accumulate when that impulse is satisfied, irrespective of the result, which in this case happens to be worthless. No mental faculty, assisted by sight, informs her when she has enough, or when she has too little. An instinctive predisposition is her only guide, an infallible guide under normal conditions, but hopelessly lost when subjected to the wiles of the experimenter. Had the Bee the least glimmer of reason would she lay her egg on the third, on the tenth part of the necessary provender? Would she lay it in an empty cell? Would she be guilty of such inconceivable maternal aberration as to leave her nurseling without nourishment? I have told the story; let the reader decide.

This instinctive predisposition, which does not leave the insect free to act and, through that very fact, saves it from error, bursts forth under yet another aspect. Let us grant the Bee as much judgment as you please. Thus endowed, will she be capable of meting out the future's larva's portion? By no means. The Bee does not know what that portion is. There is nothing to tell the materfamilias; and yet, at her first attempt, she fills the honey-pot to the requisite depth. True, in her childhood she received a similar ration, but she consumed it in the darkness of a cell; and besides, as a grub, she was blind. Sight was not her informant: it did not tell her the quantity of the provisions. Did memory, the memory of the stomach that once digested them? But digestion took place a year ago; and since that distant epoch, the nurseling, now an adult insect, has changed its shape, its dwelling, its mode of life. It was a grub; it is a Bee. Does the actual insect remember that childhood's meal? No more than we remember the sups of milk drawn from our mother's breast. The Bee, therefore, knows nothing of the quantity of provisions needed by her larva, whether from memory, from example or from acquired experience. Then what guides her when she makes her estimate with such precision? Judgment and sight would leave the mother greatly perplexed, liable to provide too much or not enough. To instruct her beyond the possibility of a mistake demands a special tendency, an unconscious impulse, an instinct, an inward voice that dictates the measure to be apportioned.

CHAPTER 8
PARASITES.

In August or September, let us go into some gorge with bare and sun- scorched sides. When we find a slope well-baked by the summer heat, a quiet corner with the temperature of an oven, we will call a halt: there is a fine harvest to be gathered there. This tropical land is the native soil of a host of Wasps and Bees, some of them busily piling the household provisions in underground warehouses: here a stack of Weevils, Locusts or Spiders, there a whole assortment of Flies, Bees, Mantes or Caterpillars, while others are storing up honey in membranous wallets or clay pots, or else in cottony bags or urns made with the punched-out disks of leaves.

With the industrious folk who go quietly about their business, the labourers, masons, foragers, warehousers, mingles the parasitic tribe, the prowlers hurrying from one home to the next, lying in wait at the doors, watching for a favourable opportunity to settle their family at the expense of others.

A heart-rending struggle, in truth, is that which rules the insect world and in a measure our own world too. No sooner has a worker, by dint of exhausting labour, amassed a fortune for his children than the non-producers come hastening up to contend for its possession. To one who amasses there are sometimes five, six or more bent upon his ruin; and often it ends not merely in robbery but in black murder. The worker's family, the object of so much care, for whom that home was built and those provisions stored, succumb, devoured by the intruders, directly the little bodies have acquired the soft roundness of youth. Shut up in a cell that is closed on every side, protected by its silken covering, the grub, once its victuals are consumed, sinks into a profound slumber, during which the organic changes needed for the future transformation take place. For this new hatching, which is to turn a grub into a Bee, for this general remodelling, the delicacy of which demands absolute repose, all the precautions that make for safety have been taken.

These precautions will be foiled. The enemy will succeed in penetrating the impregnable fortress; each foe has his special tactics, contrived with appalling skill. See, an egg is inserted by means of a probe beside the torpid larva; or else, in the absence of such an implement, an infinitesimal grub, an atom, comes creeping and crawling, slips in and reaches the sleeper, who will never wake again, already a succulent morsel for her ferocious visitor. The interloper makes the victim's cell and cocoon his own cell and his own cocoon; and next year, instead of the mistress of the house, there will come from below ground the bandit who usurped the dwelling and consumed the occupant.

Look at this one, striped black, white and red, with the figure of a clumsy, hairy Ant. She explores the slope on foot, inspects every nook and corner, sounds the soil with her antennae. She is a Mutilla, the scourge of the cradled grubs. The female has no wings, but, being a Wasp, she carries a sharp poniard. To novice eyes she would easily pass for a sort of robust Ant, distinguished from the common ruck by her garb of staring motley. The male, wide-winged and more gracefully shaped, hovers incessantly a few inches above the sandy expanse. For hours at a time, on the same spot, after the manner of the Scolia-wasp he spies the coming of the females out of the ground. If our watch be patient and persevering, we shall see the mother, after trotting about for a bit, stop somewhere and begin to scratch and dig, finally laying bare a subterranean gallery, of which there was nothing to betray the entrance; but she can discern what is invisible to us. She penetrates into the abode, remains there for a while and at last reappears to replace the rubbish and close the door as it was at the start. The abominable deed is done: the Mutilla's egg has been laid in another's cocoon, beside the slumbering larva on which the newborn grub will feed.

Here are others, all aglitter with metallic gleams: gold, emerald, blue and purple. They are the humming-birds of the insect-world, the Chrysis-wasps, or Golden Wasps, another set of exterminators of the larvae overcome with lethargy in their cocoons. In them, the atrocious assassin of cradled children lies hidden under the splendour of the garb. One of them, half emerald and half pale-pink, Parnopes carnea by name, boldly enters the burrow of Bembex rostrata at the very moment when the mother is at home, bringing a fresh piece to her larva, whom she feeds from day to day. To the elegant criminal, unskilled in navvy's work, this is the one moment to find the door open. If the mother were away, the house would be shut up; and the Golden Wasp, that sneak-thief in royal robes, could not get in. She enters, therefore, dwarf as she is, the house of the giantess whose ruin she is meditating; she makes her way right to the back, all heedless of the Bembex, her sting and her powerful jaws. What cares she that the home is not deserted? Either unmindful of the danger or paralysed with terror, the Bembex mother lets her have her way.

The unconcern of the invaded is equalled only by the boldness of the invader. Have I not seen the Anthophora-bee, at the door to her dwelling, stand a little to one side and make room for the Melecta to enter the honey-stocked cells and substitute her family for the unhappy parent's? One would think that they were two friends meeting on the threshold, one going in, the other out!

It is written in the book of fate: everything shall happen without impediment in the burrow of the Bembex; and next year, if we open the cells of that mighty huntress of Gad-flies, we shall find some which contain a russet-silk cocoon, the shape of a thimble with its orifice closed with a flat lid. In this silky tabernacle, which is protected by the hard outer shell, is a Parnopes carnea. As for the grub of the Bembex, that grub which wove the silk and next encrusted the outer casing with sand, it has disappeared entirely, all but the tattered remnants of its skin. Disappeared how? The Golden Wasp's grub has eaten it.

Another of these splendid malefactors is decked in lapis-lazuli on the thorax and in Florentine bronze and gold on the abdomen, with a terminal scarf of azure. The nomenclators have christened her Stilbum calens, FAB. When Eumenes Amedei (A species of Mason-wasp. — Translator's Note.) has built on the rock her agglomeration of dome- shaped cells, with a casing of little pebbles set in the plaster, when the store of Caterpillars is consumed and the secluded ones have hung their apartments with silk, we see the Stilbum take her stand on the inviolable citadel. No doubt some imperceptible cranny, some defect in the cement, allows her to insert her ovipositor, which shoots out like a probe. At any rate, about the end of the following May, the Eumenes' chamber contains a cocoon which again is shaped like a thimble. From this cocoon comes a Stilbum calens. There is nothing left of the Eumenes' grub: the Golden Wasp has gorged herself upon it.

Flies play no small part in this brigandage. Nor are they the least to be dreaded, weaklings though they be, sometimes so feeble that the collector dare not take them in his fingers for fear of crushing them. There are some clad in velvet so extraordinarily delicate that the least touch rubs it off. They are fluffs of down almost as frail, in their soft elegance, as the crystalline edifice of a snowflake before it touches ground. They are called Bombylii.

With this fragility of structure is combined an incomparable power of flight. See this one, hovering motionless two feet above the ground. Her wings vibrate so rapidly that they appear to be in repose. The insect looks as though it were hung at one point in space by some invisible thread. You make a movement; and the Bombylius has disappeared. You cast your eyes in search of her around you, far away, judging the distance by the vigour of her flight. There is nothing here, nothing there. Then where is she? Close by you. Look at the point whence she started: the Bombylius is there again, hovering motionless. From this aerial observatory, as quickly recovered as quitted, she inspects the ground, watching for the favourable moment to establish her egg at the cost of another creature's destruction. What does she covet for her offspring: the honey-cupboard, the stores of game, the larvae in their transformation-sleep? I do not know yet, What I do know is that her slender legs and her dainty velvet dress do not allow her to make underground searches. When she has found the propitious place, suddenly she will swoop down, lay her egg on the surface in that lightning touch with the tip of her abdomen and straightway fly up again. What I suspect, for reasons set forth presently, is that the grub that comes out of the Bombylius' egg must, of its own motion, at its own risk and peril, reach the victuals which the mother knows to be close at hand. She has no strength to do more; and it is for the new-born grub to make its way into the refectory.

I am better acquainted with the manoeuvres of certain Tachinae, the tiniest of pale-grey Flies, who, cowering on the sand in the sun, in the neighbourhood of a burrow, patiently await the hour at which to strike the fell blow. Let a Bembex-wasp return from the chase, with her Gad-fly; a Philanthus, with her Bee; a Cerceris, with her Weevil; a Tachytes, with her Locust: straightway the parasites are there, coming and going, turning and twisting with the Wasp, always at her rear, without allowing themselves to be put off by any cautious feints. At the moment when the huntress goes indoors, with her captured game between her legs, they fling themselves on her prey, which is on the point of disappearing underground, and nimbly lay their eggs upon it. The thing is done in the twinkling of an eye: before the threshold is crossed, the carcase holds the germs of a new set of guests, who will feed on victuals not amassed for them and starve the children of the house to death.

This other, resting on the burning sand, is also a member of the Fly tribe; she is an Anthrax. (Cf. "The Life of the Fly": chapter 2. — Translator's Note.) She has wide wings, spread horizontally, half smoked and half transparent. She wears a dress of velvet, like the Bombylius, her near neighbour in the official registers; but, though the soft down is similar in fineness, it is very different in colour. Anthrax is Greek for coal. It is a happy denomination, reminding us of the Fly's mourning livery, a coal-black livery with silver tears. The same deep mourning garbs those parasitic Bees, and these are the only instances known to me of that violent opposition of dead black and white.

Nowadays, when men interpret everything with glorious assurance, when they explain the Lion's tawny mane as due to the colour of the African desert, attribute the Tiger's dark stripes to the streaks of shadow cast by the bamboos and extricate any number of other magnificent things with the same facility from the mists of the unknown, I should not be sorry to hear what they have to say of the Melecta, the Crocisa and the Anthrax and of the origin of their exceptional costume.

The word 'mimesis' has been invented for the express purpose of designating the animal's supposed faculty of adapting itself to its environment by imitating the objects around it, at least in the matter of colouring. We are told that it uses this faculty to baffle its foes, or else to approach its prey without alarming it. Finding itself the better for this dissimulation, a source of prosperity indeed, each race, sifted by the struggle for life, is considered to have preserved those best-endowed with mimetic powers and to have allowed the others to become extinct, thus gradually converting into a fixed characteristic what at first was but a casual acquisition. The Lark became earth-coloured in order to hide himself from the eyes of the birds of prey when pecking in the fields; the Common Lizard adopted a grass-green tint in order to blend with the foliage of the thickets in which he lurks; the Cabbage-caterpillar guarded against the bird's beak by taking the colour of the plant on which it feeds. And so with the rest.

In my callow youth, these comparisons would have interested me: I was just ripe for that kind of science. In the evenings, on the straw of the threshing-floor, we used to talk of the Dragon, the monster which, to inveigle people and snap them up with greater certainty, became indistinguishable from a rock, the trunk of a tree, a bundle of twigs. Since those happy days of artless credulity, scepticism has chilled my imagination to some extent. By way of a parallel with the three examples which I have quoted, I ask myself why the White Wagtail, who seeks his food in the furrows as does the Lark, has a white shirt- front surmounted by a magnificent black stock. This dress is one of those most easily picked out at a distance against the rusty colour of the soil. Whence this neglect to practise mimesis, 'protective mimicry'? He has every need of it, poor fellow, quite as much as his companion in the fields!

Why is the Eyed Lizard of Provence as green as the Common Lizard, considering that he shuns verdure and chooses as his haunt, in the bright sunlight, some chink in the naked rocks where not so much as a tuft of moss grows? If, to capture his tiny prey, his brother in the copses and the hedges thought it necessary to dissemble and consequently to dye his pearl-embroidered coat, how comes it that the denizen of the sun-blistered rocks persists in his blue-and-green colouring, which at once betrays him against the whity-grey stone? Indifferent to mimicry, is he the less skilful Beetle-hunter on that account, is his race degenerating? I have studied him sufficiently to be able to declare with positive certainty that he continues to thrive both in numbers and in vigour.

Why has the Spurge-caterpillar adopted for its dress the gaudiest colours and those which contrast most with the green of the leaves which it frequents? Why does it flaunt its red, black and white in patches clashing violently with one another? Would it not be worth its while to follow the example of the Cabbage-caterpillar and imitate the verdure of the plant that feeds it? Has it no enemies? Of course it has: which of us, animals and men, has not?

A string of these whys could be extended indefinitely. It would give me amusement, did my time permit me, to counter each example of protective mimicry with a host of examples to the contrary. What manner of law is this which has at least ninety-nine exceptions in a hundred cases? Poor human nature! There is a deceptive agreement between a few actual facts and the theory which we are so foolishly ready to believe; and straightway we interpret the facts in the light of the theory. In a speck of the immense unknown we catch a glimpse of a phantom truth, a shadow, a will-o'-the-wisp; once the atom is explained, for better or worse, we imagine that we hold the explanation of the universe and all that it contains; and we forthwith shout:

'The great law of Nature! Behold the infallible law! '

Meanwhile, the discordant facts, an innumerable host, clamour at the gates of the law, being unable to gain admittance.

At the door of that infinitely restricted law clamour the great tribe of Golden Wasps, whose dazzling splendour, worthy of the wealth of Golconda, clashes with the dingy colour of their haunts. To deceive the eyes of their bird-tyrants, the Swift, the Swallow, the Chat and the others, these Chrysis-wasps, who glow like a carbuncle, like a nugget in the midst of its dark veinstone, certainly do not adapt themselves to the sand and the clay of their downs. The Green Grasshopper, we are told, thought out a plan for gulling his enemies by identifying himself in colour with the grass in which he dwells, whereas the Wasp, so rich in instinct and strategy, allowed herself to be distanced in the race by the dull-witted Locust! Rather than adapt herself as the other does, she persists in her incredible splendour, which betrays her from afar to every insect-eater and in particular to the little Grey Lizard, who lies hungrily in wait for her on the old sun-tapestried walls. She remains ruby, emerald and turquoise amidst her grey environment; and her race thrives none the worse.

The enemy that eats you is not the only one to be deceived; mimesis must also play its colour-tricks on him whom you have to eat. See the Tiger in his jungle, see the Praying Mantis on her green branch. (For the Praying Mantis, cf. "Social Life in the Insect World", by J. H. Fabre, translated by Bernard Miall: chapters 5 to 7. —Translator's Note.) Astute mimicry is even more necessary when the one to be duped is an amphitryon at whose cost the parasite's family is to be established. The Tachinae seem to declare as much: they are grey or greyish, of a colour as undecided as the dusty soil on which they cower while waiting for the arrival of the huntress laden with her capture. But they dissemble in vain: the Bembex, the Philanthus and the others see them from above, before touching ground; they recognize them perfectly at a distance, despite their grey costume. And so they hover prudently above the burrow and strive, by sudden feints, to mislead the traitorous little Fly, who, on her side, knows her business too well to allow herself to be enticed away or to leave the spot where the other is bound to return. No, a thousand times no: clay-coloured though they be, the Tachinae have no better chance of attaining their ends than a host of other parasites whose clothing is not of grey frieze to match the locality frequented, as witness the glittering Chrysis, or the Melecta and the Crocisa, with their white spots on a black ground.

We are also told that, the better to cozen his amphitryon, the parasite adopts more or less the same shape and colouring; he turns himself, in appearance, into a harmless neighbour, a worker belonging to the same guild. Instance the Psithyrus, who lives at the expense of the Bumble-bee. But in what, if you please, does Parnopes carnea resemble the Bembex into whose home she penetrates in her presence? In what does the Melecta resemble the Anthophora, who stands aside on her threshold to let her pass? The difference of costume is most striking. The Melecta's deep mourning has naught in common with the Anthophora's russet coat. The Parnopes' emerald-and-carmine thorax possesses not the least feature of resemblance with the black-and-yellow livery of the Bembex. And this Chrysis also is a dwarf in comparison with the ardent Nimrod who goes hunting Gad-flies.

Besides, what a curious idea, to make the parasite's success depend upon a more or less faithful likeness with the insect to be robbed! Why, the imitation would have exactly the opposite effect! With the exception of the Social Bees, who work at a common task, failure would be certain, for here, as among mankind, two of a trade never agree. An Osmia, an Anthophora, a Chalicodoma had better be careful not to poke an indiscreet head in at her neighbour's door: a sound drubbing would soon recall her to a sense of the proprieties. She might easily find herself with a dislocated shoulder or a mangled leg in return for a simple visit which was perhaps prompted by no evil intention. Each for herself in her own stronghold. But let a parasite appear, meditating foul play: that's a very different thing. She can wear the trappings of Harlequin or of a church-beadle; she can be the Clerus-beetle, in wing-cases of vermilion with blue trimmings, or the Dioxys-bee, with a red scarf across her black abdomen, and the mistress of the house will let her have her way, or, if she become too pressing, will drive her off with a mere flick of her wing. With her, there is no serious fray, no fierce fight. The Bludgeon is reserved for the friend of the family. Now go and practice your mimesis in order to receive a welcome from the Anthophora or the Chalicodoma! A few hours spent with the insects themselves will turn any one into a hardened scoffer at these artless theories.

To sum up, mimesis, in my eyes, is a piece of childishness. Were I not anxious to remain polite, I should say that it is sheer stupidity; and the word would express my meaning better. The variety of combinations in the domain of possible things is infinite. It is undeniable that, here and there, cases occur in which the animal harmonizes with surrounding objects. It would even be very strange if such cases were excluded from actuality, since everything is possible. But these rare coincidences are faced, under exactly similar conditions, by inconsistencies so strongly marked and so numerous that, having frequency on their side, they ought, in all logic, to serve as the basis of the law. Here, one fact says yes; there, a thousand facts say no. To which evidence shall we lend an ear? If we only wish to bolster up a theory, it would be prudent to listen to neither. The how and why escapes us; what we dignify with the pretentious title of a law is but a way of looking at things with our mind, a very squint-eyed way, which we adopt for the requirements of our case. Our would-be laws contain but an infinitesimal shade of reality; often indeed they are but puffed out with vain imaginings. Such is the law of mimesis, which explains the Green Grasshopper by the green leaves in which this Locust settles and is silent as to the Crioceris, that coral-red Beetle who lives on the no less green leaves of the lily.

And it is not only a mistaken interpretation: it is a clumsy pitfall in which novices allow themselves to be caught. Novices, did I say? The greatest experts themselves fall into the trap. One of our masters of entomology did me the honour to visit my laboratory. I was showing my collection of parasites. One of them, clad in black and yellow, attracted his attention.

'This, ' said he, 'is obviously a parasite of the Wasps. '

Surprised at the statement, I interposed:

'By what signs do you know her? '

'Why look: it's the exact colouring of the Wasp, a mixture of black and yellow. It is a most striking case of mimesis. '

'Just so; nevertheless, our black-and-yellow friend is a parasite of the Chalicodoma of the Walls, who has nothing in common, either in shape or colour, with the Wasp. This is a Leucopsis, not one of whom enters the Wasps' nest. '

'Then mimesis...? '

'Mimesis is an illusion which we should do well to relegate to oblivion. '

And, with the evidence, a whole series of conclusive examples, in front of him, my learned visitor admitted with a good grace that his first convictions were based on a most ludicrous foundation.

A piece of advice to beginners: you will go wrong a thousand times for once that you are right if, when anxious to obtain a premature sight of the probable habits of an insect, you take mimesis as your guide. With mimesis above all, it is wise, when the law says that a thing is black, first to enquire whether it does not happen to be white.

Let us go on to more serious subjects and enquire into parasitism itself, without troubling any longer about the costume of the parasite. According to etymology, a parasite is one who eats another's bread, one who lives on the provisions of others. Entomology often alters this term from its real meaning. Thus it describes as parasites the Chrysis, the Mutilla, the Anthrax, the Leucopsis, all of whom feed their family not on the provisions amassed by others, but on the very larvae which have consumed those provisions, their actual property. When the Tachinae have succeeded in laying their eggs on the game warehoused by the Bembex, the burrower's home is invaded by real parasites, in the strict sense of the word. Around the heap of Gad- flies, collected solely for the children of the house, new guests force their way, numerous and hungry, and without the least ceremony plunge into the thick of it. They sit down to a table that was not laid for them; they eat side by side with the lawful owner; and this in such haste that he dies of starvation, though he is respected by the teeth of the interlopers who have gorged themselves on his portion.

When the Melecta has substituted her egg for the Anthophora's, here again we see a real parasite settling in the usurped cell. The pile of honey laboriously gathered by the mother will not even be broken in upon by the nurseling for which it was intended. Another will profit by it, with none to say him nay. Tachinae and Melectae: those are the true parasites, consumers of others' goods.

Can we say as much of the Chrysis or the Mutilla? In no wise. The Scoliae, whose habits are known to us, are certainly not parasites. (The habits of the Scolia-wasp have been described in different essays not yet translated into English. —Translator's Note.) No one will accuse them of stealing the food of others. Zealous workers, they seek and find under ground the fat grubs on which their family will feed. They follow the chase by virtue of the same quality as the most renowned hunters, Cerceris, Sphex or Ammophila; only, instead of removing the game to a special lair, they leave it where it is, down in the burrow. Homeless poachers, they let their venison be consumed on the spot where it is caught.

In what respect do the Mutilla, the Chrysis, the Leucopsis, the Anthrax and so many others differ, in their way of living, from the Scolia? It seems to me, in none. See for yourselves. By an artifice that varies according to the mother's talent, their grubs, either in the germ-stage or newly-born, are brought into touch with the victim that is to feed them: an unwounded victim, for most of them are without a sting; a live victim, but steeped in the torpor of the coming transformations and thus delivered without defence to the grub that is to devour it.

With them, as with the Scoliae, meals are made on the spot on game legitimately acquired by indefatigable battues or by patient stalking in which all the rules have been observed; only, the animal hunted is defenceless and does not need to be laid low with a dagger-thrust. To seek and find for one's larder a torpid prey incapable of resistance is, if you like, less meritorious than heroically to stab the strong- jawed Rose-chafer or Rhinoceros-beetle; but since when has the title of sportsman been denied to him who blows out the brains of a harmless Rabbit, instead of waiting without flinching for the furious charge of the Wild Boar and driving his hunting-knife into him behind his shoulder? Besides, if the actual assault is without danger, the approach is attended with a difficulty that increases the merit of these second-rate poachers. The coveted game is invisible. It is confined in the stronghold of a cell and moreover protected by the surrounding wall of a cocoon. Of what prowess must not the mother be capable to determine the exact spot at which it lies and to lay her egg on its side or at least close by? For these reasons, I boldly number the Chrysis, the Mutilla and their rivals among the hunters and reserve the ignoble title of parasites for the Tachina, the Melecta, the Crocisa, the Meloe-beetle, in short, for all those who feed on the provisions of others.

All things considered, is ignoble the right epithet to apply to parasitism? No doubt, in the human race, the idler who feeds at other people's tables is contemptible at all points; but must the animal bear the burden of the indignation inspired by our own vices? Our parasites, our scurvy parasites, live at their neighbour's expense: the animal never; and this changes the whole aspect of the question. I know of no instance, not one, excepting man, of parasites who consume the provisions hoarded by a worker of the same species. There may be, here and there, a few cases of larceny, of casual pillage among hoarders belonging to the same trade: that I am quite ready to admit, but it does not affect things. What would be really serious and what I formally deny is that, in the same zoological species, there should be some who possessed the attribute of living at the expense of the rest. In vain do I consult my memory and my notes: my long entomological career does not furnish me with a solitary example of such a misdeed as that of an insect leading the life of a parasite upon its fellows.

When the Chalicodoma of the Sheds works, in her thousands, at her Cyclopean edifice, each has her own home, a sacred home where not one of the tumultuous swarm, except the proprietress, dreams of taking a mouthful of honey. It is as though there were a neighbourly understanding to respect the others' rights. Moreover, if some heedless one mistakes her cell and so much as alights on the rim of a cup that does not belong to her, forthwith the owner appears, admonishes her severely and soon calls her to order. But, if the store of honey is the estate of some deceased Bee, or of some wanderer unduly prolonging her absence, then—and then alone—a kinswoman seizes upon it. The goods were waste property, which she turns to account; and it is a very proper economy. The other Bees and Wasps behave likewise: never, I say never, do we find among them an idler assiduously planning the conquest of her neighbour's possessions. No insect is a parasite on its own species.

What then is parasitism, if one must look for it among animals of different races? Life in general is but a vast brigandage. Nature devours herself; matter is kept alive by passing from one stomach into another. At the banquet of life, each is in turn the guest and the dish; the eater of to-day becomes the eaten of tomorrow; hodie tibi, cras mihi. Everything lives on that which lives or has lived; everything is parasitism. Man is the great parasite, the unbridled thief of all that is fit to eat. He steals the milk from the Lamb, he steals the honey from the children of the Bee, even as the Melecta pilfers the pottage of the Anthophora's sons. The two cases are similar. Is it the vice of indolence? No, it is the fierce law which for the life of the one exacts the death of the other.

In this implacable struggle of devourers and devoured, of pillagers and pillaged, of robbers and robbed, the Melecta deserves no more than we the title of ignoble; in ruining the Anthophora, she is but imitating man in one detail, man who is the infinite source of destruction. Her parasitism is no blacker than ours: she has to feed her offspring; and, possessing no harvesting-tools, ignorant besides of the art of harvesting, she uses the provisions of others who are better endowed with implements and talents. In the fierce riot of empty bellies, she does what she can with the gifts at her disposal.

CHAPTER 9
THE THEORY OF PARASITISM.

The Melecta does what she can with the gifts at her disposal. I should leave it at that, if I had not to take into consideration a grave charge brought against her. She is accused of having lost, for want of use and through laziness, the workman's tools with which, so we are told, she was originally endowed. Finding it to her advantage to do nothing, bringing up her family free of expense, to the detriment of others, she is alleged to have gradually inspired her race with an abhorrence for work. The harvesting-tools, less and less often employed, dwindled and perished as organs having no function; the species changed into a different one; and finally idleness turned the honest worker of the outset into a parasite. This brings us to a very simple and seductive theory of parasitism, worthy to be discussed with all respect. Let us set it forth.

Some mother, nearing the end of her labours and in a hurry to lay her eggs, found, let us suppose, some convenient cells provisioned by her fellows. There was no time for nest-building and foraging; if she would save her family, she must perforce appropriate the fruit of another's toil. Thus relieved of the tedium and fatigue of work, freed of every care but that of laying eggs, she left a progeny which duly inherited the maternal slothfulness and handed this down in its turn, in a more and more accentuated form, as generation followed on generation; for the struggle for life made this expeditious way of establishing yourself one of the most favourable conditions for the success of the offspring. At the same time, the organs of work, left unemployed, became atrophied and disappeared, while certain details of shape and colouring were modified more or less, so as to adapt themselves to the new circumstances. Thus the parasitic race was definitely established.

This race, however, was not too greatly transformed for us to be able, in certain cases, to trace its origin. The parasite has retained more than one feature of those industrious ancestors. So, for instance, the Psithyrus is extremely like the Bumble-bee, whose parasite and descendant she is. The Stelis preserves the ancestral characteristics of the Anthidium; the Coelioxys-bee recalls the Leaf-cutter.

Thus speak the evolutionists, with a wealth of evidence derived not only from correspondence in general appearance, but also from similarity in the most minute particulars. Nothing is small: I am as much convinced of that as any man; and I admire the extraordinary precision of the details furnished as a basis for the theory. But am I convinced? Rightly or wrongly, my turn of mind does not hold minutiae of structure in great favour: a joint of the palpi leaves me rather cold; a tuft of bristles does not appear to me an unanswerable argument. I prefer to question the creature direct and to let it describe its passions, its mode of life, its aptitudes. Having heard its evidence, we shall see what becomes of the theory of parasitism.

Before calling upon it to speak, why should I not say what I have on my mind? And mark me, first of all, I do not like that laziness which is said to favour the animal's prosperity. I have also believed and I still persist in believing that activity alone strengthens the present and ensures the future both of animals and men. To act is to live; to work is to go forward. The energy of a race is measured by the aggregate of its action.

No, I do not like it at all, this idleness so much commended of science. We have quite enough of these zoological brutalities: man, the son of the Ape; duty, a foolish prejudice; conscience, a lure for the simple; genius, neurosis; patriotism, jingo heroics; the soul, a product of protoplasmic energies; God, a puerile myth. Let us raise the war-whoop and go out for scalps; we are here only to devour one another; the summum bonum is the Chicago packer's dollar-chest! Enough, quite enough of that, without having transformism next to break down the sacred law of work. I will not hold it responsible for our moral ruin; it has not a sturdy enough shoulder to effect such a breach; but still it has done its worst.

No, once more, I do not like those brutalities which, denying all that gives some dignity to our wretched life, stifle our horizon under an extinguisher of matter. Oh, don't come and forbid me to think, though it were but a dream, of a responsible human personality, of conscience, of duty, of the dignity of labour! Everything is linked together: if the animal is better off, as regards both itself and its race, for doing nothing and exploiting others, why should man, its descendant, show greater scruples? The principle that idleness is the mother of prosperity would carry us far indeed. I have said enough on my own account; I will call upon the animals themselves, more eloquent than I.

Are we so very sure that parasitic habits come from a love of inaction? Did the parasite become what he is because he found it excellent to do nothing? Is repose so great an advantage to him that he abjured his ancient customs in order to obtain it? Well, since I have been studying the Bee who endows her family with the property of others, I have not yet seen anything in her that points to slothfulness. On the contrary, the parasite leads a laborious life, harder than that of the worker. Watch her on a slope blistered by the sun. How busy she is, how anxious! How briskly she covers every inch of the radiant expanse, how indefatigable she is in her endless quests; in her visits, which are generally fruitless! Before coming upon a nest that suits her, she has dived a hundred times into cavities of no value, into galleries not yet victualled. And then, however kindly her host, the parasite is not always well received in the hostelry. No, it is not all roses in her trade. The expenditure of time and labour which she finds necessary in order to house an egg may easily equal or even exceed that of the worker in building her cell and filling it with honey. That industrious one has regular and continuous work, an excellent condition for success in her egg-laying; the other has a thankless and precarious task, at the mercy of a thousand accidents which endanger the great undertaking of installing the eggs. One has only to watch the prolonged hesitation of a Coelioxys seeking for the Leaf-cutters' cells to recognize that the usurpation of another's nest is not effected without serious difficulties. If she turned parasite in order to make the rearing of her offspring easier and more prosperous, certainly she was very ill-inspired. Instead of rest, hard work; instead of a flourishing family, a meagre progeny.

To generalities, which are necessarily vague, we will add some precise facts. A certain Stelis (Stelis nasuta, LATR.) is a parasite of the Mason-bee of the Walls. When the Chalicodoma has finished building her dome of cells upon her pebble, the parasite appears, makes a long inspection of the outside of the home and proposes, puny as she is, to introduce her eggs into this cement fortress. Everything is most carefully closed: a layer of rough plaster, at least two-fifths of an inch thick, entirely covers the central accumulation of cells, which are each of them sealed with a thick mortar plug. And it is the honey of these well-guarded chambers that has to be reached by piercing a wall almost as hard as rock.

The parasite pluckily sets to; the idler becomes a glutton for work. Atom by atom, she perforates the general enclosure and scoops out a shaft just sufficient for her passage; she reaches the lid of the cell and gnaws it until the coveted provisions appear in sight. It is a slow and painful process, in which the feeble Stelis wears herself out, for the mortar is much the same as Roman cement in hardness. I myself find a difficulty in breaking it with the point of my knife. What patient effort, then, the task requires from the parasite, with her tiny pincers!

I do not know exactly how long the Stelis takes to make her entrance- shaft, as I have never had the opportunity or rather the patience to follow the work from start to finish; but what I do know is that a Chalicodoma of the Walls, incomparably larger and stronger than the parasite, when demolishing before my eyes the lid of a cell sealed only the day before, was unable to complete her undertaking in one afternoon. I had to come to her assistance in order to discover, before the end of the day, the object of her housebreaking. When the Mason-bee's mortar has once set, its resistance is that of stone. Now the Stelis has not only to pierce the lid of the honey-store; she must also pierce the general casing of the nest. What a time it must take her to get through such a task, a gigantic one for her poor tools!

It is done at last, after infinite labour. The honey appears. The Stelis slips through and, on the surface of the provisions, side by side with the Chalicodoma's eggs, the number varying from time to time. The victuals will be the common property of all the new arrivals, whether the son of the house or strangers.

The violated dwelling cannot remain as it is, exposed to marauders from without; the parasite must herself wall up the breach which she has contrived. The quondam housebreaker becomes a builder. At the foot of the pebble, the Stelis collects a little of that red earth which characterizes our stony plateaus grown with lavender and thyme; she makes it into mortar by wetting it with saliva; and with the pellets thus prepared she fills up the entrance-shaft, displaying all the care and art of a regular master-mason. Only, the work clashes in colour with the Chalicodoma's. The Bee goes and gathers her cementing-powder on the adjoining high-road, the metal of which consists of broken flint-stones, and very seldom uses the red earth under the pebble supporting the nest. This choice is apparently dictated by the fact that the chemical properties of the former are more likely to produce a solid structure. The lime of the road, mixed with saliva, yields a harder cement than red clay would do. At any rate, the Chalicodoma's nest is more or less white because of the source of its materials. When a red speck, a few millimetres wide, appears on this pale background, it is a sure sign that a Stelis has been that way. Open the cell that lies under the red stain: we shall find the parasite's numerous family established there. The rusty spot is an infallible indication that the dwelling has been violated: at least, it is so in my neighbourhood, where the soil is as I have described.

We see the Stelis, therefore, at first a rabid miner, using her mandibles against the rock; next a kneader of clay and a plasterer restoring broken ceilings. Her trade does not seem one of the least arduous. Now what did she do before she took to parasitism? Judging from her appearance, the transformists tell us that she was an Anthidium, that is to say, she used to gather the soft cotton-wool from the dry stalks of the lanate plants and fashion it into wallets, in which to heap up the pollen-dust which she gleaned from the flowers by means of a brush carried on her abdomen. Or else, springing from a genus akin to the cotton-workers, she used to build resin partitions in the spiral stairway of a dead Snail. Such was the trade driven by her ancestors.

Really! So, to avoid slow and painful work, to achieve an easy life, to give herself the leisure favourable to the settlement of her family, the erstwhile cotton-presser or collector of resin-drops took to gnawing hardened cement! She who once sipped the nectar of flowers made up her mind to chew concrete! Why, the poor wretch toils at her filing like a galley-slave! She spends more time in ripping up a cell than it would take her to make a cotton wallet and fill it with food. If she really meant to progress, to do better in her own interest and that of her family, by abandoning the delicate occupations of the old days, we must confess that she has made a strange mistake. The mistake would be no greater if fingers accustomed to fancy-weaving were to lay aside velvet and silk and proceed to handle the quarryman's blocks or to break stones on the roadside.

No, the animal does not commit the folly of voluntarily embittering its lot; it does not, in obedience to the promptings of idleness, give up one condition to embrace another and a more irksome; should it blunder for once, it will not inspire its posterity with a wish to persevere in a costly delusion. No, the Stelis never abandoned the delicate art of cotton-weaving to break down walls and to grind cement, a class of work far too unattractive to efface the memory of the joys of harvesting amid the flowers. Indolence has not evolved her from an Anthidium. She has always been what she is to-day: a patient artificer in her own line, a steady worker at the task that has fallen to her share.

That hurried mother who first, in remote ages, broke into the abode of her fellows to secure a home for her eggs found this unscrupulous method, so you tell us, very favourable to the success of her race, by virtue of its economy of time and trouble. The impression left by this new policy was so profound that heredity bequeathed it to posterity, in ever-increasing proportions, until at last parasitic habits became definitely fixed. The Chalicodoma of the Sheds, followed by the Three- horned Osmia, will teach us what to think of this conjecture.

I have described in an earlier chapter my installation of Chalicodoma- hives against the walls of a porch facing the south. Here, on a level with my head, placed so that they can easily be observed, hang some tiles removed from the neighbouring roofs in winter, together with their enormous nests and their occupants. Every May, for five or six years in succession, I have assiduously watched the works of my Mason- bees. From the mass of my notes on the subject I take the following experiments which bear upon the matter under discussion.

Long ago, when I used to scatter a handful of Chalicodomae some way from home, in order to study their capacity for finding their nest again, I noticed that, if they were too long absent, the laggards found their cells closed on their return. Neighbours had taken the opportunity to lay their eggs there, after finishing the building and stocking it with provisions. The abandoned property benefited another. On realizing the usurpation, the Bee returning from her long journey soon consoled herself for the mishap. She began to break the seals of some cell or other, adjoining her own; the rest let her have her way, being doubtless too busy with their present labours to seek a quarrel with the freebooter. As soon as she had destroyed the lid, the Bee, with a sort of feverish haste that burned to repay theft by theft, did a little building, did a little victualling, as though to resume the thread of her occupations, destroyed the egg in being, laid her own and closed the cell again. Here was a touch of nature that deserved careful examination.

At eleven o'clock in the morning, when the work is at its height, I mark half-a-score of Chalicodomae with different colours, to distinguish them from one another. Some are occupied with building, others are disgorging honey. I mark the corresponding cells in the same way. As soon as the marks are quite dry, I catch the ten Bees, place them singly in screws of paper and shut them all in a box until the next morning. After twenty-four hours' captivity, the prisoners are released. During their absence, their cells have disappeared under a layer of recent structures; or, if still exposed to view, they are closed and others have made use of them.

As soon as they are free, the ten Bees, with one exception, return to their respective tiles. They do more than this, so accurate is their memory, despite the confusion resulting from a prolonged incarceration: they return to the cell which they have built, the beloved stolen cell; they minutely explore the outside of it, or at least what lies nearest to it, if the cell has disappeared under the new structures. In cases where the home is not henceforward inaccessible, it is at least occupied by a strange egg and the door is securely fastened. To this reverse of fortune the ousted ones retort with the brutal lex talionis: an egg for an egg, a cell for a cell. You've stolen my house; I'll steal yours. And, without much hesitation, they proceed to force the lid of a cell that suits them. Sometimes they recover possession of their own home, if it is possible to get into it; sometimes and more frequently they seize upon some one else's, even at a considerable distance from their original dwelling.

Patiently they gnaw the mortar lid. As the general rough-cast covering all the cells is not applied until the end of the work, all that they need do is to demolish the lid, a hard and wearisome task, but not beyond the strength of their mandibles. They therefore attack the door, the cement disk, and reduce it to dust. The criminal is allowed to carry out her nefarious designs without the slightest interference or protest from any of her neighbours, though these must necessarily include the chief party interested. The Bee is as forgetful of her cell of yesterday as she is jealous of her actual cell. To her the present is everything; the past means nothing; and the future means no more. And so the population of the tile leave the breakers of doors to do their business in peace; none hastens to the defence of a home that might well be her own. How differently things would happen if the cell were still on the stocks! But it dates back to yesterday, to the day before; and no one gives it another thought.

It's done: the lid is demolished; access is free. For some time, the Bee stands bending over the cell, her head half-buried in it, as though in contemplation. She goes away, she returns undecidedly; at last she makes up her mind. The egg is snapped up from the surface of the honey and flung on the rubbish-heap with no more ceremony than if the Bee were ridding the house of a bit of dirt. I have witnessed this hideous crime again and yet again; I confess to having repeatedly provoked it. In housing her egg, the Mason-bee displays a brutal indifference to the fate of her neighbour's egg.

I see some of them afterwards busy provisioning, disgorging honey and brushing pollen into the cell already completely provisioned; I see some masoning a little at the orifice, or at least laying on a few trowels of mortar. It seems as if the Bee, although the victuals and the building are just as they should be, were resuming the work at the point at which she left it twenty-four hours before. Lastly, the egg is laid and the opening closed up. Of my captives, one, less patient than the rest, rejects the slow process of eating away the cover and decides in favour of robbery with violence, on the principle that might is right. She dislodges the owner of a half-stocked cell, keeps good watch for a long time on the threshold of the home and, when she feels herself the mistress of the house, goes on with the provisioning. I follow the ousted proprietress with my eyes. I see her seize upon a closed cell by breaking into it, behaving in all respects like my imprisoned Chalicodomae.

The whole occurrence was too significant to be left without further confirmation. I repeated the experiment, therefore, almost every year, always with the same success. I can only add that, among the Bees placed by my artifices under the necessity of making up for lost time, a few are of a more easy-going temperament. I see some building anew, as if nothing out of the way had happened; others— this is a very rare course—going to settle on another tile, as though to avoid a society of thieves; and lastly a few who bring pellets of mortar and zealously finish the lid of their own cell, although it contains a strange egg. However, housebreaking is the usual thing.

One more detail not without value: it is not necessary for you to intervene and imprison Mason-bees for a time in order to witness the acts of violence which I have described. If you follow the work of the swarm assiduously, you may occasionally find a surprise awaiting you. A Mason-bee will appear and, for no reason known to you, break open a door and lay her egg in the violated cell. From what goes before, I look upon the Bee as a laggard, kept away from the workyard by an accident, or else carried to a distance by a gust of wind. On returning after an absence of some duration, she finds her place taken, her cell used by another. The victim of an usurper's villainy, like the prisoners in my paper screws, she behaves as they do and indemnifies herself for her loss by breaking into another's home.

Lastly, it was a matter of learning the behaviour, after their act of violence, of the Masons who have smashed in a door, brutally expelled the egg within and replaced it by one of their own laying. When the lid is repaired to look as good as new and everything restored to order, will they continue their burglarious ways and exterminate the eggs of others to make room for their own? By no means. Revenge, that pleasure of the gods and perhaps also of Bees, is satisfied after one cell has been ripped open. All anger is appeased when the egg for which so much work has been done is safely housed. Henceforth, both prisoners and stray laggards resume their ordinary labours, indifferently with the rest. They build honestly, they provision honestly, nor meditate further evil. The past is quite forgotten until a fresh disaster occurs.

To return to the parasites: a mother chanced to find herself the mistress of another's nest. She took advantage of this to entrust her egg to it. This expeditious method, so easy for the mother and so favourable to the success of her offspring, made such an impression on her that she transmitted the maternal indolence to her posterity. Thus the worker gradually became transformed into a parasite.

Capital! The thing goes like clockwork, as long as we have only to put our ideas on paper. But let us just consult the facts, if you don't mind; before arguing about probabilities, let us look into things as they are. Here is the Mason-bee of the Sheds teaching us something very curious. To smash the lid of a cell that does not belong to her, to throw the egg out of doors and put her own in its place is a practice which she has followed since time began. There is no need of my interference to make her commit burglary: she commits it of her own accord, when her rights are prejudiced as the result of a too-long absence. Ever since her race has been kneading cement, she has known the law of retaliation. Countless ages, such as the evolutionists require, have made her adopt forcible usurpation as an inveterate habit. Moreover, robbery is so incomparably easy for the mother. No more cement to scratch up with her mandibles on the hard ground, no more mortar to knead, no more clay walls to build, no more pollen to gather on hundreds and hundreds of journeys. All is ready, board and lodging. Never was a better opportunity for allowing one's self a good time. There is nothing against it. The others, the workers, are imperturbable in their good-humour. Their outraged cells leave them profoundly indifferent. There are no brawls to fear, no protests. Now or never is the moment to tread the primrose path.

Besides, your progeny will be all the better for it. You can choose the warmest and wholesomest spots; you can multiply your laying-operations by devoting to them all the time that you would have to spend on irksome occupations. If the impression produced by the violent seizure of another's property is strong enough to be handed down by heredity, how deep should be the impression of the actual moment when the Mason-bee is in the first flush of success! The precious advantage is fresh in the memory, dating from that very instant; the mother has but to continue in order to create a method of installation favourable in the highest degree to her and hers. Come, poor Bee! Throw aside your exhausting labours, follow the evolutionists' advice and, as you have the means at your disposal, become a parasite!

But no, having effected her little revenge, the builder returns to her masonry, the gleaner to her gleaning, with unquenchable zeal. She forgets the crime committed in a moment of anger and takes good care not to hand down any tendency towards idleness to her offspring. She knows too well that activity is life, that work is the world's great joy. What myriads of cells has she not broken open since she has been building; what magnificent opportunities, all so clear and conclusive, has she not had to emancipate herself from drudgery! Nothing could convince her: born to work, she persists in an industrious life. She might at least have produced an offshoot, a race of housebreakers, who would invade cells by demolishing doors. The Stelis does something of the kind; but who would think of proclaiming a relationship between the Chalicodoma and her? The two have nothing in common. I call for a scion of the Mason-bee of the Sheds who shall live by the art of breaking through ceilings. Until they show me one, the theorists will only make me smile when they talk to me of erstwhile workers relinquishing their trade to become parasitic sluggards.

I also call, with no less insistence, for a descendant of the Three-horned Osmia, a descendant given to demolishing party-walls. I will describe later how I managed to make a whole swarm of these Osmiae build their nests on the table in my study, in glass tubes that enabled me to see the inmost secrets of the work of the Bee. (Cf.

"Bramble-bees and Others", by J. Henri Fabre, translated by Alexander Teixeira de Mattos: chapters 1 to 7. —Translator's Note.) For three or four weeks, each Osmia is scrupulously faithful to her tube, which is laboriously filled with a set of chambers divided by earthen partitions. Marks of different colours painted on the thorax of the workers enable me to recognize individuals in the crowd. Each crystal gallery is the exclusive property of one Osmia; no other enters it, builds in it or hoards in it. If, through heedlessness, through momentary forgetfulness of her own house in the tumult of the city, some neighbour so much as comes and looks in at the door, the owner soon puts her to flight. No such indiscretion is tolerated. Every Bee has her home and every home its Bee.

All goes well until just before the end of the work. The tubes are then closed at the orifice with a thick plug of earth; nearly the whole swarm has disappeared; there remain on the spot a score of tatterdemalions in threadbare fleeces, worn out by a month's hard toil. These laggards have not finished their laying. There is no lack of unoccupied tubes, for I take care to remove some of those which are full and to replace them by others that have not yet been used. Very few of the Bees decide to take possession of these new homes, which differ in no particular from the earlier ones; and even then they build only a small number of cells, which are often mere attempts at partitions.

They want something different: a nest belonging to some one else. They bore through the stopper of the inhabited tubes, a work of no great difficulty, for we have here not the hard cement of the Chalicodoma, but a simple lid of dried mud. When the entrance is cleared, a cell appears, with its store of provisions and its egg, with her brutal mandibles; she rips it open and goes and flings it away. She does worse: she eats it on the spot. I had to witness this horror many times over before I could accept it as a fact. Note that the egg devoured may very well contain the criminal's own offspring. Imperiously swayed by the needs of her present family, the Osmia puts her past family entirely out of her mind.

Having perpetrated this child-murder, the depraved creature does a little provisioning. They all experience the same necessity to go backwards in the sequence of actions in order to pick up the thread of their interrupted occupations. Her next work is to lay her egg and then she conscientiously restores the demolished lid.

The havoc can be more sweeping still. One of these laggards is not satisfied with a single cell; she needs two, three, four. To reach the most remote, the Osmia wrecks all those which come before it. The partitions are broken down, the eggs eaten or thrown away, the provisions swept outside and often even carried to a distance in great lumps. Covered with dust from the loose plaster of the demolition, floured all over with the rifled pollen, sticky with the contents of the mangled eggs, the Osmia, while at her brigand's work, is altered beyond recognition. Once the place is cleared, everything resumes its normal course. Provisions are laboriously brought to take the place of those which have been thrown away; eggs are laid, one on each heap of food; the partitions are built up again; and the massive plug sealing the whole structure is made as good as new.

Crimes of this kind recur so often that I am obliged to interfere and place in safety the nests which I wish to keep intact. And nothing as yet explains this brigandage, bursting forth at the end of the work like a moral epidemic, like a frenzied delirium. I should say nothing if the site were lacking; but the tubes are there, close by, empty and quite fit to receive the eggs. The Osmia refuses them, she prefers to plunder. Is it from weariness, from a distaste for work after a period of fierce activity? Not at all; for, when a row of cells has been stripped of its contents, after the ravage and waste, she has to come back to ordinary work, with all its burdens. The labour is not reduced; it is increased. It would pay the Bee infinitely better, if she wants to continue her laying, to make her home in an unoccupied tube. The Osmia thinks differently. Her reasons for acting as she does escape me. Can there be ill-conditioned characters among her, characters that delight in a neighbour's ruin? There are among men.

In the privacy of her native haunts, the Osmia, I have no doubt, behaves as in my crystal galleries. Towards the end of the building-operations, she violates others' dwellings. By keeping to the first cell, which it is not necessary to empty in order to reach the next, she can utilize the provisions on the spot and shorten to that extent the longest part of her work. As usurpations of this kind have had ample time to become inveterate, to become inbred in the race, I ask for a descendant of the Osmia who eats her grandmother's egg in order to establish her own egg.

This descendant I shall not be shown; but I may be told that she is in process of formation. The outrages which I have described are preparing a future parasite. The transformists dogmatize about the past and dogmatize about the future, but as seldom as possible talk to us about the present. Transformations have taken place, transformations will take place; the pity of it is that they are not actually taking place. Of the three tenses, one is lacking, the very one which directly interests us and which alone is clear of the incubus of theory. This silence about the present does not please me overmuch, scarcely more than the famous picture of "The Crossing of the Red Sea" painted for a village chapel. The artist had put upon the canvas a broad ribbon of brightest scarlet; and that was all.

'Yes, that's the Red Sea, ' said the priest, examining the masterpiece before paying for it. 'That's the Red Sea, right enough; but where are the Israelites? '

'They have passed, ' replied the painter.

'And the Egyptians? '

'They are on the way. '

Transformations have passed, transformations are on the way. For mercy's sake, cannot they show us transformations in the act? Must the facts of the past and the facts of the future necessarily exclude the facts of the present? I fail to understand.

I call for a descendant of the Chalicodoma and a descendant of the Osmia who have robbed their neighbours with gusto, when occasion offered, since the origin of their respective races, and who are working industriously to create a parasite happy in doing nothing. Have they succeeded? No. Will they succeed? Yes, people maintain. For the moment, nothing. The Osmiae and Chalicodomae of to-day are what they were when the first trowel of cement or mud was mixed. Then how many ages does it take to form a parasite? Too many, I fear, for us not to be discouraged.

If the sayings of the theorists are well-founded, going on strike and living by shifts was not always enough to assure parasitism. In certain cases, the animal must have had to change its diet, to pass from live prey to vegetarian fare, which would entirely subvert its most essential characteristics. What should we say to the Wolf giving up mutton and browsing on grass, in obedience to the dictates of idleness? The boldest would shrink from such an absurd assumption. And yet transformism leads us straight to it.

Here is an example: in July, I split some bramble-stems in which Osmia tridentata has built her nests. In the long series of cells, the lower already hold the Osmia's cocoons, while the upper contain the larva which has nearly finished consuming its provisions and the topmost show the victuals untouched, with the Osmia's egg upon them. It is a cylindrical egg, rounded at both extremities, of a transparent white and measuring four to five millimetres in length. (. 156 to. 195 inch. - -Translator's Note.) It lies slantwise, one end of it resting on the food and the other sticking up at some distance above the honey. Now, by multiplying my visits to the fresh cells, I have on several occasions made a very valuable discovery. On the free end of the Osmia's egg, another egg is fixed; an egg quite different in shape, white and transparent like the first, but much smaller and narrower, blunt at one end and tapering into a rather sharp point at the other. It is two millimetres long by half a millimetre wide. (. 078 and. 019 inch. —Translator's Note.) It is undeniably the egg of a parasite, a parasite which compels my attention by its curious method of installing its family.

It opens before the Osmia's egg. The tiny grub, as soon as it is born, begins to drain the rival egg, of which it occupied the top part, high up above the honey. The extermination soon becomes perceptible. You can see the Osmia's egg turning muddy, losing its brilliancy, becoming limp and wrinkled. In twenty-four hours, it is nothing but an empty sheath, a crumpled bit of skin. All competition is now removed; the parasite is the master of the house. The young grub, when demolishing the egg, was active enough: it explored the dangerous thing which had to be got rid of quickly, it raised its head to select and multiply the attacking-points. Now, lying at full length on the surface of the honey, it no longer shifts its position; but the undulations of the digestive canal betray its greedy absorption of the Osmia's store of food. The provisions are finished in a fortnight and the cocoon is woven. It is a fairly firm ovoid, of a very dark-brown colour, two characteristics which at once distinguish it from the Osmia's pale, cylindrical cocoon. The hatching takes place in April or May. The puzzle is solved at last: the Osmia's parasite is a Wasp called the Spotted Sapyga (Sapyga punctata, V.L.)

Now where are we to class this Wasp, a true parasite in the strict sense of the word, that is to say, a consumer of others' provisions.

Her general appearance and her structure make it clear to any eye more or less familiar with entomological shapes that she belongs to a species akin to that of the Scoliae. Moreover, the masters of classification, so scrupulous in their comparison of characteristics, agree in placing the Sapygae immediately after the Scoliae and a little before the Mutillae. The Scoliae feed their grubs on prey; so do the Mutillae. The Osmia's parasite, therefore, if it really derives from a transformed ancestor, is descended from a flesh-eater, though it is now an eater of honey. The Wolf does more than become a Sheep: he turns himself into a sweet-tooth.

'You will never get an apple-tree out of an acorn, ' Franklin tells us, with that homely common-sense of his.

In this case, the passion for jam must have sprung from a love of venison. Any theory might well be deficient in balance when it leads to such vagaries as this.

I should have to write a volume if I would go on setting forth my doubts. I have said enough for the moment. Man, the insatiable enquirer, hands down from age to age his questions about the whys and wherefores of origins. Answer follows answer, is proclaimed true to-day and recognized as false tomorrow; and the goddess Isis continues veiled.

CHAPTER 10
THE TRIBULATIONS OF THE MASON-BEE.

To illustrate the methods of those who batten on others' goods, the plunderers who know no rest till they have wrought the destruction of the worker, it would be difficult to find a better instance than the tribulations suffered by the Chalicodoma of the Walls. The Mason who builds on the pebbles may fairly boast of being an industrious workwoman. Throughout the month of May, we see her black squads, in the full heat of the sun, digging with busy teeth in the mortar-quarry of the road hard by. So great is her zeal that she hardly moves out of the way of the passer-by; more than one allows herself to be crushed underfoot, absorbed as she is in collecting her cement.

The hardest and driest spots, which still retain the compactness imparted by the steam-roller, are the favourite veins; and the work of making the pellet is slow and painful. It is scraped up atom by atom; and, by means of saliva, turned into mortar then and there. When it is all well kneaded and there is enough to make a load, the Mason sets off with an impetuous flight, in a straight line, and makes for her pebble, a few hundred paces away. The trowel of fresh mortar is soon spent, either in adding another storey to the turret-shaped edifice, or in cementing into the wall lumps of gravel that give it greater solidity. The journeys in search of cement are renewed until the structure attains the regulation height. Without a moment's rest, the Bee returns a hundred times to the stone-yard, always to the one spot recognized as excellent.

The victuals are now collected: honey and flower-dust. If there is a pink carpet of sainfoin anywhere in the neighbourhood, 'tis there that the Mason goes plundering by preference, though it cost her a four hundred yards' journey every time. Her crop swells with honeyed exudations, her belly is floured with pollen. Back to the cell, which slowly fills; and back straightway to the harvest-field. And all day long, with not a sign of weariness, the same activity is maintained as long as the sun is high enough. When it is late, if the house is not yet closed, the Bee retires to her cell to spend the night there, head downwards, tip of her abdomen outside, a habit foreign to the Chalicodoma of the Sheds. Then and then alone the Mason rests; but it is a rest that is in a sense equivalent to work, for, thus placed, she blocks the entrance to the honey-store and defends her treasure against twilight or night marauders.

Being anxious to form some estimate of the total distance covered by the Bee in the construction and provisioning of a single cell, I counted the number of steps from a nest to the road where the mortar was mixed and from the same nest to the sainfoin-field where the harvest was gathered. I took such note as my patience permitted of the journeys made in both directions; and, completing these data with a comparison between the work done and that which remained to do, I arrived at nine and a half miles as the result of the total travelling. Of course, I give this figure only as a rough calculation; greater precision would have demanded more perseverance than I can boast.

Such as it is, the result, which is probably under the actual figure in many cases, is of a kind that gives us a vivid idea of the Mason- bee's activity. The complete nest will comprise about fifteen cells. Moreover, the heap of cells will be coated at the end with a layer of cement a good finger's-breadth thick. This massive fortification, which is less finished than the rest of the work but more expensive in materials, represents perhaps in itself one half of the complete task, so that, to establish her dome, Chalicodoma muraria, coming and going across the arid table-land, traverses altogether a distance of 275 miles, which is nearly half of the greatest dimension of France from north to south. Afterwards, when, worn out with all this fatigue, the Bee retires to a hiding-place to languish in solitude and die, she is surely entitled to say:

'I have laboured, I have done my duty! '

Yes, certainly, the Mason has toiled with a vengeance. To ensure the future of her offspring, she has spent her own life without reserve, her long life of five or six weeks' duration; and now she breathes her last, contented because everything is in order in the beloved house: copious rations of the first quality; a shelter against the winter frosts; ramparts against incursions of the enemy. Everything is in order, at least so she thinks; but, alas, what a mistake the poor mother is making! Here the hateful fatality stands revealed, aspera fata, which ruins the producer to provide a living for the drone; here we see the stupid and ferocious law that sacrifices the worker for the idler's benefit. What have we done, we and the insects, to be ground with sovran indifference under the mill-stone of such wretchedness? Oh, what terrible, what heart-rending questions the Mason-bee's misfortunes would bring to my lips, if I gave free scope to my sombre thoughts! But let us avoid these useless whys and keep within the province of the mere recorder.

There are some ten of them plotting the ruin of the peaceable and industrious Bee; and I do not know them all. Each has her own tricks, her own art of injury, her own exterminating tactics, so that no part of the Mason's work may escape destruction. Some seize upon the victuals, others feed on the larvae, others again convert the dwelling to their own use. Everything has to submit: cell, provisions, scarce- weaned nurselings.

The stealers of food are the Stelis-wasp (Stelis nasuta) and the Dioxys-bee (Dioxys cincta). I have already said how, in the Mason's absence, the Stelis perforates the dome of cell after cell, lays her eggs there and afterwards repairs the breach with a mortar made of red earth, which at once betrays the parasite's presence to a watchful eye. The Stelis, who is much smaller than the Chalicodoma, finds enough food in a single cell for the rearing of several of her grubs. The mother lays a number of eggs, which I have seen vary between the extremes of two and twelve, on the surface, next to the Mason's egg, which itself undergoes no outrage whatever.

Things do not go so badly at first. The feasters swim—it is the only word—in the midst of plenty; they eat and digest like brothers. Presently, times become hard for the hostess' son; the food decreases, dearth sets in; and at length not an atom remains, although the Mason's larva has attained at most a quarter of its growth. The others, more expeditious feeders, have exhausted the victuals long before the victim has finished his normal repast. The swindled grub shrivels up and dies, while the gorged larvae of the Stelis begin to spin their strong little brown cocoons, pressed close together and lumped into one mass, so as to make the best use of the scanty space in the crowded dwelling. Should you inspect the cell later, you will find, between the heaped cocoons on the wall, a little dried-up corpse. It is the larva that was such an object of care to the mother Mason. The efforts of the most laborious of lives have ended in this lamentable relic. It has happened to me just as often, when examining the secrets of the cell which is at once cradle and tomb, not to come upon the deceased grub at all. I picture the Stelis, before laying her own eggs, destroying the Chalicodoma's egg and eating it, as the Osmiae do among themselves; or I picture the dying thing, an irksome mass for the numerous spinners at work in a narrow habitation, being cut to pieces to make room for the medley of cocoons. But to so many deeds of darkness I would not like to add another by an oversight; and I prefer to admit that I failed to perceive the grub that died of hunger.

Let us now show up the Dioxys. At the time when the work of construction is in progress, she is an impudent visitor of the nests, exploiting with the same effrontery the enormous cities of the Mason- bee of the Sheds and the solitary cupolas of the Mason-bee of the Pebbles. An innumerable population, coming and going, humming and buzzing, strikes her with no awe. On the tiles hanging from the walls of my porch I see her, with her red scarf round her body, stalking with sublime assurance over the ridged expanse of nests. Her black schemes leave the swarm profoundly indifferent; not one of the workers dreams of chasing her off, unless she should come bothering too closely. Even then, all that happens is a few signs of impatience on the part of the hustled Bee. There is no serious excitement, no eager pursuits such as the presence of a mortal enemy might lead us to suspect. They are there in their thousands, each armed with her dagger; any one of them is capable of slaying the traitress; and not one attacks her. The danger is not suspected.

Meanwhile, she inspects the workyard, moves freely among the ranks of the Masons and bides her time. If the owner be absent, I see her diving into a cell, coming out again a moment later with her mouth smeared with pollen. She has been to try the provisions. A dainty connoisseur, she goes from one store to another, taking a mouthful of honey. Is it a tithe for her personal maintenance, or a sample tested for the benefit of her coming grub? I should not like to say. What I do know is that, after a certain number of these tastings, I catch her stopping in a cell, with her abdomen at the bottom and her head at the orifice. This is the moment of laying, unless I am much mistaken.

When the parasite is gone, I inspect the home. I see nothing abnormal on the surface of the mass. The sharper eye of the owner, when she gets back, sees nothing either, for she continues the victualling without betraying the least uneasiness. A strange egg, laid on the provisions, would not escape her. I know how clean she keeps her warehouse; I know how scrupulously she casts out anything introduced by my agency: an egg that is not hers, a bit of straw, a grain of dust. So, according to my evidence and that of the

Chalicodoma, which is more conclusive, the Dioxys's egg, if it is really laid then, is not placed on the surface.

I suspect, without having yet verified my suspicion—and I reproach myself for the neglect—I suspect that the egg is buried in the heap of pollen-dust. When I see the Dioxys come out of a cell with her mouth all over yellow flour, perhaps she has been surveying the ground and preparing a hiding-place for her egg. What I take for a mere tasting might well be a more serious act. Thus concealed, the egg escapes the eagle eye of the Bee, whereas, if left uncovered, it would inevitably perish, would be flung on the rubbish heap at once by the owner of the nest. When the Spotted Sapyga lays her egg on that of the Bramble-dwelling Osmia, she does the deed under cover of darkness, in the gloom of a deep well to which not the least ray of light can penetrate; and the mother, returning with her pellet of green putty to build the closing partition, does not see the usurping germ and is ignorant of the danger. But here everything happens in broad daylight; and this demands more cunning in the method of installation.

Besides, it is the one favourable moment for the Dioxys. If she waits for the Mason-bee to lay, it is too late, for the parasite is not able to break down doors, as the Stelis does. As soon as her egg is laid, the Mason-bee of the Sheds comes out of her cell and at once turns round and proceeds to close it up with the pellet of mortar which she holds ready in her mandibles. The material is employed with such method that the actual sealing is done in a moment: the other pellets, the object of repeated journeys, will serve merely to increase the thickness of the lid. The chamber is inaccessible to the Dioxys from the first touch of the trowel. Hence it is absolutely necessary for her to see to her egg before the Mason-bee of the Sheds has disposed of hers and no less necessary to conceal it from the Mason's watchful eye.

The difficulties are not so great in the nests of the Mason-bee of the Pebbles. After this Bee has laid her egg, she leaves it for a time to go in search of the cement needed for closing the cell; or, if she already holds a pellet in her mandibles, this is not enough to seal it properly, as the orifice is larger. More pellets are needed to wall up the entrance entirely. The Dioxys would have time to strike her blow during the mother's absences; but everything seems to suggest that she behaves on the pebbles as she does on the tiles. She steals a march by hiding the egg in the mass of pollen and honey.

What becomes of the Mason's egg confined in the same cell with the egg of the Dioxys? In vain have I opened nests at every season; I have never found a vestige of the egg nor of the grub of either Chalicodoma. The Dioxys, whether as a larva on the honey, or enclosed in its cocoon, or as the perfect insect, was always alone. The rival had disappeared without a trace. A suspicion thereupon suggests itself; and the facts are so compelling that the suspicion is almost equal to a certainty. The parasitic grub, which hatches earlier than the other, emerges from its hiding-place, from the midst of the honey, comes to the surface and, with its first bite, destroys the egg of the Mason-bee, as the Sapyga does the egg of the Osmia. It is an odious, but a supremely efficacious method. Nor must we cry out too loudly against such foul play on the part of a new born infant: we shall meet with even more heinous tactics later. The criminal records of life are full of these horrors which we dare not search too deeply. An infinitesimal creature, a barely-visible grub, with the swaddling- clothes of its egg still clinging to it, is led by instinct, at its first inspiration, to exterminate whatever is in its way.

So the Mason's egg is exterminated. Was it really necessary in the Dioxys' interest? Not in the least. The hoard of provisions is too large for its requirements in a cell of the Chalicodoma of the Sheds; how much more so in a cell of the Chalicodoma of the Pebbles! She eats not a half, hardly a third of it. The rest remains as it was, untouched. We see here, in the destruction of the Mason's egg, a flagrant waste which aggravates the crime. Hunger excuses many things; for lack of food, the survivors on the raft of the Medusa indulged in a little cannibalism; but here there is enough food and to spare. When there is more than she needs, what earthly motive impels the Dioxys to destroy a rival in the germ stage? Why cannot she allow the larva, her mess-mate, to take advantage of the remains and afterwards to shift for itself as best it can? But no: the Mason-bee's offspring must needs be stupidly sacrificed on the top of provisions which will only grow mouldy and useless! I should be reduced to the gloomy lucubrations of a Schopenhauer if I once let myself begin on parasitism.

Such is a brief sketch of the two parasites of the Chalicodoma of the Pebbles, true parasites, consumers of provisions hoarded on behalf of others. Their crimes are not the bitterest tribulations of the Mason-bee. If the first starves the Mason's grub to death, if the second makes it perish in the egg, there are others who have a more pitiable ending in store for the worker's family. When the Bee's grub, all plump and fat and greasy, has finished its provisions and spun its cocoon wherein to sleep the slumber akin to death, the necessary period of preparation for its future life, these other enemies hasten to the nests whose fortifications are powerless against their hideously ingenious methods. Soon on the sleeper's body lies a nascent grub which feasts in all security on the luscious fare. The traitors who attack the larvae in their lethargy are three in number: an Anthrax, a Leucopsis and a microscopic dagger-wearer. (Monodontomerus cupreus. For this and the Anthrax, cf. "The Life of the Fly": chapters 2 and 3. The Leucopsis is a Hymenopteron, the essay upon whom forms the concluding chapter of the present volume. —Translator's Note.) Their story deserves to be told without reticence; and I shall tell it later. For the moment, I merely mention the names of the three exterminators.

The provisions are stolen, the egg is destroyed. The young grub dies of hunger, the larva is devoured. Is that all? Not yet. The worker must be exploited thoroughly, in her work as well as in her family. Here are some now who covet her dwelling. When the Mason is constructing a new edifice on a pebble, her almost constant presence is enough to keep the aspirants to free lodgings at a distance; her strength and vigilance overawe whoso would annex her masonry. If, in her absence, one greatly daring thinks of visiting the building, the owner soon appears upon the scene and ousts her with the most discouraging animosity. She has no need then to fear the entrance of unwelcome tenants while the house is new. But the Bee of the Pebbles also uses old dwellings for her laying, as long as they are not too much dilapidated. In the early stages of the work, neighbours compete for these with an eagerness which shows the value attached to them. Face to face, at times with their mandibles interlocked, now both rising into the air, now coming down again, then touching ground and rolling over each other, next flying up again, for hours on end they will wage battle for the property at issue.

A ready-made nest, a family heirloom which needs but a little restoring, is a precious thing for the Mason, ever sparing of her time. We find so many of the old homes repaired and restocked that I suspect the Bee of laying new foundations only when there are no secondhand nests to be had. To have the chambers of a dome occupied by a stranger therefore means a serious privation.

Now several Bees, however industrious in gathering honey, building party-walls and contriving receptacles for provisions, are less clever at preparing the resorts in which the cells are to be stacked. The abandoned chambers of the Chalicodoma, now larger than they were originally, through the addition of the hall of exit, are first-rate acquisitions for them. The great thing is to occupy these chambers first, for here possession is nine parts of the law. Once established, the Mason is not disturbed in her home, while she, in her turn, does not disturb the stranger who has settled down before her in an old nest, the patrimony of her family. The disinherited one leaves the Bohemian to enjoy the ruined manor in peace and goes to another pebble to establish herself at fresh expense.

In the first rank of these free tenants, I will place an Osmia (Osmia cyanoxantha, PEREZ) and a Megachile, or Leaf-cutting Bee (Megachile apicalis, SPIN.) (Cf. "Bramble-dwellers and Others": chapter 8. — Translator's Note.), both of whom work in May, at the same time as the Mason, while both are small enough to lodge from five to eight cells in a single chamber of the Chalicodoma, a chamber increased by the addition of an outer hall. The Osmia subdivides this space into very irregular compartments by means of slanting, upright or curved partitions, subject to the dictates of space. There is no art, consequently, in the accumulation of little cells; the architect's only task is to use the breadth at her disposal in a frugal manner. The material employed for the partitions is a green, vegetable putty, which the Osmia must obtain by chewing the shredded leaves of a plant whose nature is still uncertain. The same green paste serves for the thick plug that closes the abode. But in this case the insect does not use it unadulterated. To give greater power of resistance to the work, it mixes a number of bits of gravel with the vegetable cement. These materials, which are easily picked up, are lavishly employed, as though the mother feared lest she should not fortify sufficiently the entrance to her dwelling. They form a sort of coarse stucco, on the more or less smooth cupola of the Chalicodoma; and this unevenness, as well as the green colouring of its mortar of masticated leaves, at once betrays the Osmia's nest. In course of time, under the prolonged action of the air, the vegetable putty turns brown and assumes a dead- leaf tint, especially on the outside of the plug; and it would then be difficult for any one who had not seen them when freshly made to recognize their nature.

The old nests on the pebbles seem to suit other Osmiae. My notes mention Osmia Morawitzi, PEREZ, and Osmia cyanea, KIRB., as having been recognized in these dwellings, although they are not very assiduous visitors. Lastly, to complete the enumeration of the

Bees known to me as making their homes in the Mason's cupolas, I must add Megachile apicalis, who piles in each cell a half-dozen or more honey- pots constructed with disks cut from the leaves of the wild rose, and an Anthidium whose species I cannot state, having seen nothing of her but her white cotton sacks.

The Mason-bee of the Sheds, on the other hand, supplies free lodgings to two species of Osmiae, Osmia tricornis, LATR., and Osmia Latreillii, SPIN., both of whom are quite common. The Three-horned Osmia frequents by preference the habitations of the Bees that build their nests in populous colonies, such as the Chalicodoma of the Sheds and the Hairy-footed Anthophora. Latreille's Osmia is nearly always found with the Three-horned Osmia at the Chalicodoma's.

The real builder of the city and the exploiter of the labour of others work together, at the same period, form a common swarm and live in perfect harmony, each Bee of the two species attending to her business in peace. They share and share alike, as though by tacit agreement. Is the Osmia discreet enough not to put upon the good-natured Mason and to utilize only abandoned passages and waste cells? Or does she take possession of the home of which the real owners could themselves have made use? I lean in favour of usurpation, for it is not rare to see the Chalicodoma of the Sheds clearing out old cells and using them as does her sister of the Pebbles. Be this as it may, all this little busy world lives without strife, some building anew, others dividing up the old dwelling.

Those Osmiae, on the contrary, who are the self-invited guests of the Mason-bee of the Pebbles are the sole occupants of the dome. The cause of this isolation lies in the unsociable temper of the proprietress. The old nest does not suit her from the moment that she sees it occupied by another. Instead of going shares, she prefers to seek elsewhere a dwelling where she can work in solitude. Her gracious surrender of a most excellent lodging in favour of a stranger who would be incapable of offering the least resistance if a dispute arose proves the great immunity enjoyed by the Osmia in the home of the worker whom she exploits. The common and peaceful swarming of the Mason-bee of the Sheds and the two cell-borrowing Osmiae proves it in a still more positive fashion. There is never a fight for the acquisition of another's goods or the defence of one's own property; never a brawl between Osmiae and Chalicodomae. Robber and robbed live on the most neighbourly terms. The Osmia considers herself at home; and the other does nothing to undeceive her. If the parasites, so deadly to the workers, move about in their very ranks with impunity, without arousing the faintest excitement, an equally complete indifference must be shown by the dispossessed owners to the presence of the usurpers in their old homes. I should be greatly put to it if I were asked to reconcile this calmness on the part of the expropriated one with the ruthless competition that is said to sway the world. Fashioned so as to instal herself in the Mason's property, the Osmia meets with a peaceful reception from her. My feeble eyes can see no further.

I have named the provision-thieves, the grub-murderers and the house- grabbers who levy tribute on the Mason-bee. Does that end the list? Not at all. The old nests are cities of the dead. They contain Bees who, on achieving the perfect state, were unable to open the exit-door through the cement and who withered in their cells; they contain dead larvae, turned into black, brittle cylinders; untouched provisions, both mouldy and fresh, on which the egg has come to grief; tattered cocoons; shreds of skins; relics of the transformation.

If we remove the nest of the Chalicodoma of the Sheds from its tile — a nest sometimes quite eight inches thick — we find live inhabitants only in a thin outer layer. All the remainder, the catacombs of past generations, is but a horrible heap of dead, shrivelled, ruined, decomposed things. Into this sub-stratum of the ancient city the unreleased Bees, the untransformed larvae fall as dust; here the honey-stores of old go sour, here the uneaten provisions are reduced to mould.

Three undertakers, all members of the Beetle tribe, a Clerus, a Ptinus and an Anthrenus, batten on these remains. The larvae of the Anthrenus and the Ptinus gnaw the ashes of the corpses; the larva of the Clerus, with the black head and the rest of its body a pretty pink, appeared to me to be breaking into the old jam-pots filled with rancid honey. The perfect insect itself, garbed in vermilion with blue ornaments, is fairly common on the surface of the clay slabs during the working season, strolling leisurely through the yard to taste here and there the drops of honey oozing from some cracked pot. Notwithstanding his showy livery, so unlike the workers' sombre frieze, the Chalicodomae leave him in peace, as though they recognized in him the scavenger whose duty it is to keep the sewers wholesome.

Ravaged by the passing years, the Mason's home at last falls into ruin and becomes a hovel. Exposed as it is to the direct action of wind and weather, the dome built upon a pebble chips and cracks. To repair it would be too irksome, nor would that restore the original solidity of the shaky foundation. Better protected by the covering of a roof, the city of the sheds resists longer, without however escaping eventual decay. The storeys which each generation adds to those in which it was born increase the thickness and the weight of the edifice in alarming proportions. The moisture of the tile filters into the oldest layers, wrecks the foundations and threatens the nest with a speedy fall. It is time to abandon for good the house with its cracks and rents.

Thereupon the crumbling apartments, on the pebble as well as on the tile, become the home of a camp of gypsies who are not particular where they find a shelter. The shapeless hovel, reduced to a fragment of a wall, finds occupants, for the Mason's work must be exploited to the utmost limits of possibility. In the blind alleys, all that remains of the former cells, Spiders weave a white-satin screen, behind which they lie in wait for the passing game. In nooks which they repair in summary fashion with earthen embankments or clay partitions, Hunting Wasps—Pompili and Tripoxyla—store up small members of the Spider tribe, including sometimes the Weaving Spiders who live in the same ruins.

I have said nothing yet of the Chalicodoma of the Shrubs. My silence is not due to negligence, but to the circumstance that I am almost destitute of facts relating to her parasites. Of the many nests which I have opened in order to study their inhabitants, only one so far has been invaded by strangers. This nest, the size of a large walnut, was fixed on a pomegranate-branch. It comprised eight cells, of which seven were occupied by the Chalicodoma, and the eighth by a little Chalcis, the plague of a whole host of the Bee-tribe. Apart from this instance, which was not a very serious case, I have seen nothing. In those aerial nests, swinging at the end of a twig, not a Dioxys, a Stelis, an Anthrax, a Leucopsis, those dread ravagers of the other two Masons; never any Osmiae, Megachiles or Anthidia, those lodgers in the old buildings.

The absence of the latter is easily explained. The Chalicodoma's masonry does not last long on its frail support. The winter winds, when the shelter of the foliage has disappeared, must easily break the twig, which is little thicker than a straw and liable to give way by reason of its heavy burden. Threatened with an early fall, if it is not already on the ground, last year's dwelling is not restored to serve the needs of the present generation. The same nest does not serve twice; and this does away with the Osmiae and with their rivals in the art of utilizing old cells.

The elucidation of this point does not remove the obscurity of the next. I can see nothing to account for the absence or at least the extreme rareness of usurpers of provisions and consumers of grubs, both of whom are very indifferent to the new or old conditions of the nest, so long as the cells are well stocked. Can it be that the lofty position of the edifice and the shaky support of the twig arouse distrust in the Dioxys and other malefactors? For lack of a better explanation, I will leave it at that.

If my idea is not an empty fancy, we must admit that the Chalicodoma of the Shrubs was singularly well-inspired in building in mid-air. You have seen of what misfortunes the other two are victims. If I take a census of the population of a tile, many a time I find the Dioxys and the Mason-bee in almost equal proportions. The parasite has wiped out half the colony. To complete the disaster, it is not unusual for the grub-eaters, the Leucopsis and her rival, the pygmy Chalcis, to have decimated the other half. I say nothing of Anthrax sinuata, whom I sometimes see coming from the nests of the Chalicodoma of the Sheds; her larva preys on the Three-horned Osmia, the Mason-bee's visitor.

All solitary though she be on her boulder, which would seem the proper thing to keep away exploiters, the scourge of dense populations, the Chalicodoma of the Pebbles is no less sorely tried. My notes abound in cases such as the following: of the nine cells in one dome, three are occupied by the Anthrax, two by the Leucopsis, two by the Stelis, one by the Chalcis and the ninth by the Mason. It is as though the four miscreants had joined forces for the massacre: the whole of the Bee's family has disappeared, all but one young mother saved from the disaster by her position in the centre of the citadel. I have sometimes stuffed my pockets with nests removed from their pebbles without finding a single one that has not been violated by one or other of the malefactors and oftener still by several of them at a time. It is almost an event for me to find a nest intact. After these funereal records, I am haunted by a gloomy thought: the weal of one means the woe of another.

CHAPTER 11
THE LEUCOPSES.

(This chapter should be read in conjunction with the essays entitled
"The Anthrax" and "Larval Dimorphism", forming chapters 2 and 4
of "The Life of the Fly. " — Translator's Note.)

Let us visit the nests of Chalicodoma muraria in July, detaching them
from their pebbles with a sideward blow, as I explained when telling
the story of the Anthrax. The Mason-bee's cocoons with two
inhabitants, one devouring, the other in process of being devoured,
are numerous enough to allow me to gather some dozens in the
course of a morning, before the sun becomes unbearably hot. We will
give a smart tap to the flints so as to loosen the clay domes, wrap
these up in newspapers, fill our box and go home as fast as we can,
for the air will soon be as fiery as the devil's kitchen.

Inspection, which is easier in the shade indoors, soon tells us that,
though the devoured is always the wretched Mason-bee, the
devourer belongs to two different species. In the one case, the
cylindrical form, the creamy-white colouring and the little nipple
constituting the head reveal to us the larva of the Anthrax, which
does not concern us at present; in the other, the general structure and
appearance betray the grub of some Hymenopteron. The Mason's
second exterminator is, in fact, a Leucopsis (Leucopsis gigas, FAB.),
a magnificent insect, stripped black and yellow, with an abdomen
rounded at the end and hollowed out, as is also the back, into a
groove to contain a long rapier, as slender as a horsehair, which the
creature unsheathes and drives through the mortar right into the cell
where it proposes to establish its egg. Before occupying ourselves
with its capacities as an inoculator, let us learn how its larva lives in
the invaded cell.

It is a hairless, legless, sightless grub, easily confused, by
inexperienced eyes, with those of various honey-gathering
Hymenoptera. Its more apparent characteristics consist of a
colouring like that of rancid butter, a shiny and as it were oily skin
and a segmentation accentuated by a series of marked swellings, so
that, when looked at from the side, the back is very plainly indented.
When at rest, the larva is like a bow bending round at one point. It is
made up of thirteen segments, including the head. This head, which
is very small compared with the rest of the body, displays no mouth-

part under the lens; at most you see a faint red streak, which calls for the microscope. You then distinguish two delicate mandibles, very short and fashioned into a sharp point. A small round mouth, with a fine piercer on the right and left, is all that the powerful instrument reveals. As for my best single magnifying-glasses, they show me nothing at all. On the other hand, we can quite easily, without arming the eye with a lens, perceive the mouth-apparatus—and particularly the mandibles—of either a honey-eater, such as an Osmia, Chalicodoma or Megachile, or a game-eater, such as a Scolia, Ammophila or Bembex. All these possess stout pincers, capable of gripping, grinding and tearing. Then what is the purpose of the Leucopsis' invisible implements? His method of consuming will tell us.

Like his prototype, the Anthrax, the Leucopsis does not eat the Chalicodoma-grub, that is to say, he does not break it up into mouthfuls; he drains it without opening it and digging into its vitals. In him again we see exemplified that marvellous art which consists in feeding on the victim without killing it until the meal is over, so as always to have a portion of fresh meat. With its mouth assiduously applied to the unhappy creature's skin, the lethal grub fills itself and waxes fat, while the fostering larva collapses and shrivels, retaining just enough life, however, to resist decomposition. All that remains of the decanted corpse is the skin, which, when softened in water and blown out, swells into a balloon without the least escape of gas, thus proving the continuity of the integument. All the same, the apparently unpunctured bladder has lost its contents. It is a repetition of what the Anthrax has shown us, with this difference, that the Leucopsis seems not so well skilled in the delicate work of absorbing the victim. Instead of the clean white granule which is the sole residue when the Fly has finished her joint, the insect with the long probe has a plateful of leavings, not seldom soiled with the brownish tinge of food that has gone bad. It would seem that, towards the end, the act of consumption becomes more savage and does not disdain dead meat. I also notice that the Leucopsis is not able to get up from dinner or to sit down to it again as readily as the Anthrax. I have sometimes to tease him with the point of a hair-pencil in order to make him let go; and, once he has left the joint, he hesitates a little before putting his mouth to it again. His adhesion is not the mere result of a kiss like that of a cupping- glass; it can only be explained by hooks that need releasing.

I now see the use of the microscopic mandibles. Those two delicate spikes are incapable of chewing anything, but they may very well serve to pierce the epidermis with an aperture smaller than that made by the finest needle; and it is through this puncture that the Leucopsis sucks the juices of his prey. They are instruments made to perforate the bag of fat which slowly, without suffering any internal injury, is emptied through an opening repeated here and there. The Anthrax' cupping-glass is here replaced by piercers of exceeding sharpness and so short that they cannot hurt anything beyond the skin. Thus do we see in operation, with a different sort of implements, that wise system which keeps the provisions fresh for the consumer.

It is hardly necessary to say, to those who have read the story of the Anthrax, that this kind of feeding would be impossible with a victim whose tissues possessed their final hardness. The Mason-bee's grub is therefore emptied by the Leucopsis' larva while it is in a semifluid state and deep in the torpor of the nymphosis. The last fortnight in July and the first fortnight in August are the best times to witness the repast, which I have seen going on for twelve and fourteen days. Later, we find nothing in the Mason-bee's cocoon except the Leucopsis' larva, gloriously fat, and, by its side, a sort of thin, rancid rasher, the remains of the deceased wet-nurse. Things then remain as they are until the hot part of the following summer or at least until the end of June.

Then appears the nymph, which teaches us nothing striking; and at last the perfect insect, whose hatching may be delayed until August. Its exit from the Mason's fortress has no likeness to the strange method employed by the Anthrax. Endowed with stout mandibles, the perfect insect splits the ceiling of its abode by itself without much difficulty. At the time of its deliverance, the Mason-bees, who work in May, have long disappeared. The nests on the pebbles are all closed, the provisioning is finished, the larvae are sleeping in their yellow cocoons. As the old nests are utilized by the Mason so long as they are not too much dilapidated, the dome which has just been vacated by the Leucopsis, now more than a year old, has its other cells occupied by the Bee's children. There is here, without seeking farther, a fat living for the Leucopsis' offspring which she well knows how to turn to profit. It depends but on herself to make the house in which she was born into the residence of her family. Besides, if she has a fancy for distant exploration, clay domes abound in the harmas. The inoculation of the eggs through the walls will begin shortly. Before witnessing this curious performance, let us examine the needle that is to effect it.

The insect's abdomen is hollowed, at the top, into a furrow that runs up to the base of the thorax; the end, which is broader and rounded, has a narrow slit, which seems to divide this region into two. The whole thing suggests a pulley with a fine groove. When at rest, the inoculating-needle or ovipositor remains packed in the slit and the furrow. The delicate instrument thus almost completely encircles the abdomen. Underneath, on the median line, we see a long, dark-brown scale, pointed, keel-shaped, fixed by its base to the first abdominal segment, with its sides prolonged into membranous wings which are fastened tightly to the insect's flanks. Its function is to protect the underlying region, a soft-walled region in which the probe has its source. It is a cuirass, a lid which protects the delicate motor- machinery during periods of inactivity but swings from back to front and lifts when the implement has to be unsheathed and used.

We will now remove this lid with the scissors, so as to have the whole apparatus before our eyes, and then raise the ovipositor with the point of a needle. The part that runs along the back comes loose without the slightest difficulty, but the part embedded in the groove at the end of the abdomen offers a resistance that warns us of a complication which we did not notice at first. The tool, in fact, consists of three pieces, a central piece, or inoculating-filament, and two side-pieces, which together constitute a scabbard. The two latter are more substantial, are hollowed out like the sides of a groove and, when uniting, form a complete groove in which the filament is sheathed. This bivalvular scabbard adheres loosely to the dorsal part; but, farther on, at the tip of the abdomen and under the belly, it can no longer be detached, as its valves are welded to the abdominal wall. Here, therefore, we find, between the two joined protecting parts, a simple trench in which the filament lies covered up. As for this filament, it is easily extracted from its sheath and released down to its base, under the shield formed by the scale.

Seen under the magnifying-glass, it is a round, stiff, horny thread, midway in thickness between a human hair and a horse-hair. Its tip is a little rough, pointed and bevelled to some length down. The microscope becomes necessary if we would see its real structure, which is much less simple than it at first appears. We perceive that the bevelled end-part consists of a series of truncated cones, fitting one into the other, with their wide base slightly projecting. This arrangement produces a sort of file, a sort of rasp with very much blunted teeth. When pressed on the slide, the thread divides into four pieces of unequal length. The two longer end in the toothed bevel. They come together in a very narrow groove, which receives the two other, rather shorter pieces. These both end in a point, which, however, is not toothed and does not project as far as the final rasp. They also unite to form a groove, which fits into the groove of the other two, the whole constituting a complete channel or duct. Moreover, the two shorter pieces, considered together, can move, lengthwise, in the groove that receives them; they can also move one over the other, always lengthwise, so much so that, on the slide of the microscope, their terminal points are seldom situated on the same level.

If with our scissors we cut a piece of the inoculating-thread from the living insect and examine the section under the magnifying-glass, we shall see the inner groove lengthen out and project beyond the outer groove and then go in again in turn, while from the wound there oozes a tiny albiminous drop, doubtless proceeding from the liquid that gives the egg the singular appendage to which we shall come presently. By means of these longitudinal movements of the inner trench inside the outer trench and of the sliding, one over the other, of the two portions of the former, the egg can be despatched to the end of the ovipositor notwithstanding the absence of any muscular contraction, which is impossible in a horny conduit.

We have only to press the upper surface of the abdomen to see it disjoint itself from the first segment, as though the insect had been cut almost in two at that point. A wide gap or hiatus appears between the first and second rings; and, under a thin membrane, the base of the ovipositor bulges out, bent back into a stout hook. Here the filament passes through the insect from end to end and emerges underneath. Its issue is therefore near the base of the abdomen, instead of at the tip, as usual. This curious arrangement has the effect of shortening the lever-arm of the ovipositor and bringing the starting-point of the filament nearer to the fulcrum, namely, the legs of the insect, and of thus assisting the difficult task of inoculation by making the most of the effort expended.

To sum up, the ovipositor when at rest goes round the abdomen. Starting at the base, on the lower surface, it runs round the belly from front to back and then returns from back to front on the upper surface, where it ends at almost the same level as its starting-point. Its length is 14 millimetres. (. 546 inch—Translator's Note.) This fixes the limit of the depth which the probe is able to reach in the Mason-bee's nests.

One last word on the Leucopsis' weapon. In the dying insect, beheaded, stripped of legs and wings, with a pin stuck through its body, the sides of the fissure containing the inoculating-thread quiver violently, as if the belly were going to open, divide in two along the median line and then reunite its two halves. The thread itself gives convulsive tremblings; it comes out of its scabbard, goes back and slips out again. It is as though the laying-implement could not persuade itself to die before accomplishing its mission. The insect's supreme aim is the egg; and, so long as the least spark of life remains, it makes dying efforts to lay.

Leucopsis gigas exploits the nests of the Mason-bee of the Pebbles and the Mason-bee of the Sheds with equal zest. To observe the insertion of the egg at my ease and to watch the operator at work over and over again, I gave the preference to the last-named Mason, whose nests, removed from the neighbouring roofs by my orders, have hung for some years in the arch of my basement. These clay hives fastened to tiles supply me with fresh records each summer. I am much indebted to them in the matter of the Leucopsis' life-history.

By way of comparison with what took place under my roof, I used to observe the same scenes on the pebbles of the surrounding wastelands. My excursions, alas, did not all reward my zeal, which zeal was not without merit in the merciless sunshine; but still, at rare intervals, I succeeded in seeing some Leucopsis digging her probe into the mortar dome. Lying flat on the ground, from the beginning to the end of the operation, which sometimes lasted for hours, I closely watched the insect in its every movement, while my Dog, weary of being out of doors in that scorching heat, would discreetly retire from the fray and, with his tail between his legs and his tongue hanging out, go home and stretch himself at full length on the cool tiles of the hall. How wise he was to scorn this pebble-gazing! I would come in half- roasted, as brown as a berry, to find my friend Bull wedged into a corner, his back to the wall, sprawling on all fours, while, with heaving sides, he panted forth the last sprays of steam from his overheated interior. Yes, he was much better-advised to return as fast as he could to the shade of the house. Why does man want to know things? Why is he not indifferent to them, with the lofty philosophy of the animals? What interest can anything have for us that does not fill our stomachs? What is the use of learning? What is the use of truth, when profit is all that matters? Why am I—the descendant, so they tell me, of some tertiary Baboon—afflicted with the passion for knowledge from which Bull, my friend and companion, is exempt? Why... oh, where have I got to? I was going in, wasn't I, with a splitting headache? Quick, let us get back to our subject!

It was in the first week of July that I saw the inoculation begin on my Chalicodoma sicula nests. The parasite is at her task in the hottest part of the day, close on three o'clock in the afternoon; and work goes on almost to the end of the month, decreasing gradually in activity. I count as many as twelve Leucopses at a time on the most thickly-populated pair of tiles. The insect slowly and awkwardly explores the nests. It feels the surface with its antennae, which are bent at a right angle after the first joint. Then, motionless, with lowered head, it seems to meditate and to debate within itself on the fitness of the spot. Is it here or somewhere else that the coveted larva lies? There is nothing outside, absolutely nothing, to tell us. It is a stony expanse, bumpy but yet very uniform in appearance, for the cells have disappeared under a layer of plaster, a work of public interest to which the whole swarm devotes its last days. If I myself, with my long experience, had to decide upon the suitable point, even if I were at liberty to make use of a lens for examining the mortar grain by grain and to auscultate the surface in order to gather information from the sound emitted, I should decline the job, persuaded in advance that I should fail nine times out of ten and only succeed by chance.

Where my discernment, aided by reason and my optical contrivances, fails, the insect, guided by the wands of its antennae, never blunders. Its choice is made. See it unsheathing its long instrument. The probe points normally towards the surface and occupies nearly the central spot between the two middle-legs. A wide dislocation appears on the back, between the first and second segments of the abdomen; and the base of the instrument swells like a bladder through this opening; while the point strives to penetrate the hard clay. The amount of energy expended is shown by the way in which the bladder quivers. At every moment we expect to see the frail membrane burst with the violence of the effort. But it does not give way; and the wire goes deeper and deeper.

Raising itself high on its legs, to give free play to its apparatus, the insect remains motionless, the only sign of its arduous labours being a slight vibration. I see some perforators who have finished operating in a quarter of an hour. These are the quickest at the business. They have been lucky enough to come across a wall which is less thick and less hard than usual. I see others who spend as many as three hours on a single operation, three long hours of patient watching for me, in my anxiety to follow the whole performance to the end, three long hours of immobility for the insect, which is even more anxious to make sure of board and lodging for its egg. But then is it not a task of the utmost difficulty to introduce a hair into the thickness of a stone? To us, with all the dexterity of our fingers, it would be impossible; to the insect, which simply pushes with its belly, it is just hard work.

Notwithstanding the resistance of the substance traversed, the Leucopsis perseveres, certain of succeeding; and she does succeed, although I am still unable to understand her success. The material through which the probe has to penetrate is not a porous substance; it is homogeneous and compact, like our hardened cement. In vain do I direct my attention to the exact point where the instrument is at work; I see no fissure, no opening that can facilitate access. A miner's drill penetrates the rock only by pulverizing it. This method is not admissible here; the extreme delicacy of the implement is opposed to it. The frail stem requires, so it seems to me, a ready- made way, a crevice through which it can slip; but this crevice I have never been able to discover. What about a dissolving fluid which would soften the mortar under the point of the ovipositor? No, for I see not a trace of humidity around the point where the thread is at work. I fall back upon a fissure, a lack of continuity somewhere, although my examination fails to discover any on the Mason-bee's nest. I was better served in another case. Leucopsis dorsigera, FAB., settles her eggs on the larva of the Diadem Anthidium, who sometimes makes her nest in reed-stumps. I have repeatedly seen her insert her auger through a slight rupture in the side of the reed. As the wall was different, wood in the latter case and mortar in the former, perhaps it will be best to look upon the matter as a mystery.

My sedulous attendance, during the best part of July, in front of the tiles hanging from the walls of the arch, allowed me to reckon the inoculations. Each time that the insect, on finishing the operation, removed its probe, I marked in pencil the exact point at which the instrument was withdrawn; and I wrote down the date beside it.

This information was to be utilized when the Leucopsis finished her labours.

When the perforators are gone, I proceed with my examination of the nests, covered with my hieroglyphics, the pencilled notes. One result, one which I fully expected, compensates me straightway for all my weary waitings. Under each spot marked in black, under each spot whence I saw the ovipositor withdrawn, I always find a cell, with not a single exception. And yet there are intervals of solid stone between the cells: the partition-walls alone would account for some. Moreover, the compartments, which are very irregularly disposed by a swarm of toilers who all work in their own sweet way, have great irregular cavities between them, which end by being filled up with the general plastering of the nest. The result of this arrangement is that the massive portions cover almost the same space as the hollow portions. There is nothing outside to show whether the underlying regions are full or empty. It is quite impossible for me to decide if, by digging straight down, I shall come to a hollow cell or to a solid wall.

But the insect makes no mistake: the excavations under my pencil-marks bear witness to that; it always directs its apparatus towards the hollow of a cell. How is it apprised whether the part below is empty or full? Its organs of information are undoubtedly the antennae, which feel the ground. They are two fingers of unparalleled delicacy, which pry into the basement by tapping on the part above it. Then what do those puzzling organs perceive? A smell? Not at all; I always had my doubts of that and now I am certain of the contrary, after what I shall describe in a moment. Do they perceive a sound? Are we to treat them as a superior kind of microphone, capable of collecting the infinitesimal echoes of what is full and the reverberations of what is empty? It is an attractive idea, but unfortunately the antennae play their part equally well on a host of occasions when there are no vaults to reverberate. We know nothing and are perhaps destined never to know anything of the real value of the antennal sense, to which we have nothing analogous; but, though it is impossible for us to say what it does perceive, we are at least able to recognize to some extent what it does not perceive and, in particular, to deny it the faculty of smell.

As a matter of fact, I notice, with extreme surprise, that the great majority of the cells visited by the Leucopsis' probe do not contain the one thing which the insect is seeking, namely, the young larva of the Mason-bee enclosed in its cocoon. Their contents consist of the refuse so often met with in old Chalicodoma-nests: liquid honey left unemployed, because the egg has perished; spoilt provisions, sometimes mildewed, or sometimes a tarry mass; a dead larva, stiffened into a brown cylinder; the shrivelled corpse of a perfect insect, which lacked the strength to effect its deliverance; dust and rubbish which has come from the exit-window afterwards closed up by the outer coating of plaster. The odoriferous effluvia that can emanate from these relics certainly possess very diverse characters. A sense of smell with any subtlety at all would not be deceived by this stuff, sour, 'high, ' musty or tarry as the case may be; each compartment, according to its contents, has a special aroma, which we might or might not be able to perceive; and this aroma most certainly bears no resemblance to that which we may assume the much-desired fresh larva to possess. If nevertheless the Leucopsis does not distinguish between these various cells and drives the probe into all of them indifferently, is this not an evident proof that smell is no guide whatever to her in her search? Other considerations, when I was treating of the Hairy Ammophila, enabled me to assert that the antennae have no olfactory powers. To-day, the frequent mistakes of the Leucopsis, whose antennae are nevertheless constantly exploring the surface, make this conclusion absolutely certain.

The perforator of clay nests has, so it seems to me, delivered us from an old physiological fallacy. She would deserve studying, if for no other result than this; but her interest is far from being exhausted. Let us look at her from another point of view, whose full importance will not be apparent until the end; let us speak of something which I was very far from suspecting when I was so assiduously watching the nests of my Mason-bees.

The same cell can receive the Leucopsis' probe a number of times, at intervals of several days. I have said how I used to mark in black the exact place at which the laying-implement had entered and how I wrote the date of the operation beside it. Well, at many of these already visited spots, concerning which I possessed the most authentic documents, I saw the insect return a second, a third and even a fourth time, either on the same day or some while after, and drive its inoculating-thread in again, at precisely the same place, as though nothing had happened. Was it the same individual repeating her operation in a cell which she had visited before but forgotten, or different individuals coming one after the other to lay an egg in a compartment thought to be unoccupied? I cannot say, having neglected to mark the operators, for fear of disturbing them.

As there is nothing, except the mark of my pencil, a mark devoid of meaning to the insect, to indicate that the auger has already been at work there, it may easily happen that the same operator, finding under her feet a spot already exploited by herself but effaced from her memory, repeats the thrust of her tool in a compartment which she believes herself to be discovering for the first time. However retentive its memory for places may be, we cannot admit that the insect remembers for weeks on end, as well as point by point, the topography of a nest covering a surface of some square yards. Its recollections, if it have any, serve it badly; the outward appearance gives it no information; and its drill enters wherever it may happen to discover a cell, at points that have already perhaps been pierced several times over.

It may also happen—and this appears to me the most frequent case— that one exploiter of a cell is succeeded by a second, a third, a fourth and others still, all fired with the newcomer's zeal because their predecessors have left no trace of their passage. In one way or another, the same cell is exposed to manifold layings, though its contents, the Chalicodoma-grub, be only the bare ration of a single Leucopsis-grub.

These reiterated borings are not at all rare: I noted a score of them on my tiles; and, in the case of some cells, the operation was repeated before my eyes as often as four times. Nothing tells us that this number was not exceeded in my absence. The little that I observed prevents me from fixing any limit. And now a momentous question arises: is the egg really laid each time that the probe enters a cell? I can see not the slightest excuse for supposing the contrary. The ovipositor, because of its horny nature, can have but a very dull sense of touch. The insect is apprised of the contents of the cell only by the end of that long horse-hair, a not very trustworthy witness, I should imagine. The absence of resistance tells it that it has reached an empty space; and this is probably the only information that the insensible implement can supply. The drill boring through the rock cannot tell the miner anything about the contents of the cavern which it has entered; and the case must be the same with the rigid filament of the Leucopses.

Now that the thread has reached its goal, what does the cell contain? Mildewed honey, dust and rubbish, a shrivelled larva, or a larva in good condition? Above all, does it already contain an egg? This last question calls for a definite answer, but as a matter of fact it is impossible for the insect to learn anything from a horse-hair on that most delicate matter, the presence or absence of an egg, a mere atom of a thing, in that vast apartment. Even admitting some sense of touch at the end of the drill, one insuperable difficulty would always remain: that of finding the exact spot where the tiny speck lies in those spacious and mysterious regions. I go so far as to believe that the ovipositor tells the insect nothing, or at any rate very little, of the inside of the cell, whether propitious or not to the development of the germ. Perhaps each thrust of the instrument, provided that it meets with no resistance from solid matter, lays the egg, to whose lot there falls at one time good, wholesome food, at another mere refuse.

These anomalies call for more conclusive proofs than the rough deductions drawn from the nature of the horny ovipositor. We must ascertain in a direct fashion whether the cell into which the auger has been driven several times over actually contains several occupants in addition to the larva of the Mason-bee. When the Leucopses had finished their borings, I waited a few days longer so as to give the young grubs time to develop a little, which would make my examination easier. I then moved the tiles to the table in my study, in order to investigate their secrets with the most scrupulous care. And here such a disappointment as I have rarely known awaited me. The cells which I had seen, actually seen, with my own eyes, pierced by the probe two or three or even four times, contained but one Leucopsis-grub, one alone, eating away at its Chalicodoma. Others, which had also been repeatedly probed, contained spoilt remnants, but never a Leucopsis. O holy patience, give me the courage to begin again! Dispel the darkness and deliver me from doubt!

I begin again. The Leucopsis-grub is familiar to me; I can recognize it, without the possibility of a mistake, in the nests of both the Chalicodoma of the Pebbles and the Chalicodoma of the Sheds. All through the winter, I rush about, getting my nests from the roofs of old sheds and the pebbles of the waste-lands; I stuff my pockets with them, fill my box, load Favier's knapsack; I collect enough to litter all the tables in my study; and, when it is too cold out of doors, when the biting mistral blows, I tear open the fine silk of the cocoons to discover the inhabitant. Most of them contain the Mason in the perfect state; others give me the larva of the Anthrax; others— very numerous, these—give me the larva of the Leucopsis. And this last is alone, always alone, invariably alone. The whole thing is utterly incomprehensible when one knows, as I know, how many times the probe entered those cells.

My perplexity only increases when, on the return of summer, I witness for the second time the Leucopsis' repeated operations on the same cells and for the second time find a single larva in the compartments which have been bored several times over. Shall I then be forced to accept that the auger is able to recognize the cells already containing an egg and that it thenceforth refrains from laying there? Must I admit an extraordinary sense of touch in that bit of horse- hair, or even better, a sort of divination which declares where the egg lies without having to touch it? But I am raving! There is certainly something that escapes me; and the obscurity of the problem is simply due to my incomplete information. O patience, supreme virtue of the observer, come to my aid once more! I must begin all over again for the third time.

Until now, my investigations have been made some time after the laying, at a period when the larva is at least fairly developed. Who knows? Something perhaps happens, at the very commencement of infancy, that may mislead me afterwards. I must apply to the egg itself if I would learn the secret which the grub will not reveal. I therefore resume my observations in the first fortnight of July, when the Leucopses are beginning to visit busily both Mason-bee's nests. The pebbles in the waste-lands supply me with plenty of buildings of the Chalicodoma of the Walls; the byres scattered here and there in the fields give me, under their dilapidated roofs, in fragments broken off with the chisel, the edifices of the Chalicodoma of the Sheds. I am anxious not to complete the destruction of my home hives, already so sorely tried by my experiments; they have taught me much and can teach me more. Alien colonies, picked up more or less everywhere, provide me with my booty. With my lens in one hand and my forceps in the other, I go through my collection on the same day, with the prudence and care which only the laboratory-table permits. The results at first fall far short of my expectations. I see nothing that I have not seen before. I make fresh expeditions, after a few days' interval; I bring back fresh loads of lumps of mortar, until at last fortune smiles upon me.

Reason was not at fault. Each thrust means the laying of an egg when the probe reaches the cell. Here is a cocoon of the Mason-bee of the Pebbles with an egg side by side with the Chalicodoma-grub. But what a curious egg! Never have my eyes beheld the like; and then is it really the egg of the Leucopsis? Great was my apprehension. But I breathed again when I found, a couple of weeks later, that the egg had become the larva with which I was familiar. Those cocoons with a single egg are as numerous as I can wish; they exceed my wishes: my little glass receptacles are too few to hold them.

And here are others, more precious ones still, with manifold layings. I find plenty with two eggs; I find some with three or four; the best-colonised offer me as many as five. And, to crown my delight, the joy of the seeker to whom success comes at the last moment, when he is on the verge of despair, here again, duly furnished with an egg, is a sterile cocoon, that is to say, one containing only a shrivelled and decaying larva. All my suspicions are confirmed, down to the most inconsequent: the egg housed with a mass of putrefaction.

The nests of the Mason-bee of the Walls are the more regular in structure and are easier to examine, because their base is wide open once it is separated from the supporting pebble; and it was these which supplied me with by far the greater part of my information. Those of the Mason-bee of the Sheds have to be chipped away with a hammer before one can inspect their cells, which are heaped up anyhow; and they do not lend themselves anything like so well to delicate investigations, as they suffer both from the shock and the ill-treatment.

And now the thing is done: it remains certain that the Leucopsis' laying is exposed to very exceptional dangers. She can entrust the egg to sterile cells, without provisions fit to use; she can establish several in the same cell, though this cell contains nourishment for one only. Whether they proceed from a single individual returning several times, by inadvertence, to the same place, or are the work of different individuals unaware of the previous borings, those multiple layings are very frequent, almost as much so as the normal layings. The largest which I have noticed consisted of five eggs, but we have no authority for looking upon this number as an outside limit. Who could say, when the perforators are numerous, to what lengths this accumulation can go? I will set forth on some future occasion how the ration of one egg remains in reality the ration of one egg, despite the multiplicity of banqueters.

I will end by describing the egg, which is a white, opaque object, shaped like a much-elongated oval. One of the ends is lengthened out into a neck or pedicle, which is as long as the egg proper. This neck is somewhat wrinkled, sinuous and as a rule considerably curved. The whole thing is not at all unlike certain gourds with an elongated paunch and a snake-like neck. The total length, pedicle and all, is about 3 millimetres. (About one-eighth of an inch. — Translator's Note.) It is needless to say, after recognizing the grub's manner of feeding, that this egg is not laid inside the fostering larva. Yet, before I knew the habits of the Leucopsis, I would readily have believed that every Hymenopteron armed with a long probe inserts her eggs into the victim's sides, as the Ichneumon-flies do to the Caterpillars. I mention this for the benefit of any who may be under the same erroneous impression.

The Leucopsis' egg is not even laid upon the Mason-bee's larva; it is hung by its bent pedicle to the fibrous wall of the cocoon. When I go to work very delicately, so as not to disturb the arrangement in knocking the nest off its support, and then extract and open the cocoon, I see the egg swinging from the silken vault. But it takes very little to make it fall. And so, most often, even though it be merely the effect of the shock sustained when the nest is removed from its pebble, I find the egg detached from its suspension-point and lying beside the larva, to which it never adheres in any circumstances. The Leucopsis' probe does not penetrate beyond the cocoon traversed; and the egg remains fastened to the ceiling, in the crook of some silky thread, by means of its hooked pedicle.

Printed in the United States
71457LV00004B/205